EARTHWATCH

Earthcycles and Ecosystems

EARTHWATCH
Earthcycles and Ecosystems

Written by **Beth Savan, Ph.D.**
Illustrated by **Pat Cupples**

Addison-Wesley Publishing Company, Inc.

Reading, Massachusetts • Menlo Park, California • New York
Don Mills, Ontario • Wokingham, England • Amsterdam • Bonn
Sydney • Singapore • Tokyo • Madrid • San Juan
Paris • Seoul • Milan • Mexico City • Taipei

Library of Congress Cataloging-in-Publication Data

Savan, Beth.
    Earthwatch : earthcycles and ecosystems / written by Beth Savan ; illustrated by Pat Cupples.
           p.   cm.
    Includes index.
    Summary: Examines how natural cycles and specific ecosystems work and suggests activities to protect the Earth from further damage by pollution and waste.
    ISBN 0-201-58148-5
    1. Biotic communities — Juvenile literature.   2. Ecology — Study and teaching — Activity programs — Juvenile literature. [1. Ecology. 2. Biotic communities. 3. Pollution. 4. Conservation of natural resources. 5. Science projects.] I. Cupples, Pat, ill. II. Title.
QH541.14.S28   1992
574.5 — dc20   91-33373
                                            CIP
                                            AC

First published as *Earthcycles and Ecosystems* by Kids Can Press, Ltd., of Toronto.

Edited by Valerie Wyatt
Designed by N. R. Jackson
Set in 11-point ITC Cheltenham Book by Techni Process Limited

1 2 3 4 5 6 7 8 9-AL-95949392
First printing, January 1992

Addison-Wesley books are available at special discounts for bulk purchases by schools, institutions, and other organizations. For more information, please contact:

Special Markets Department
Addison-Wesley Publishing Company
Reading, MA 01867
(617) 944-3700 x 2431

 Text stock contains over 50% recycled paper

To my children,
Daniel, Laura and Anna

Acknowledgements

I would like to thank the Canada Council for their generous funding. Many thanks also to Mary Margaret Crapper, Sheila McAllister and Edna Lim for research; and to Beverley Hanna-Thorpe and countless others for their advice and assistance. I am especially grateful to my editor, Val Wyatt, for her editing skills, ideas and support.

CONTENTS

CHAPTER 1:
What are earthcycles?

You may not realize it, but you're part of a cycle. This cycle goes round and round like the wheels on your bicycle. The sun starts the cycle spinning.

This "earthcycle" is actually made up of several smaller cycles— the water cycle, the air cycle and the soil cycle. This book is about these cycles. It's also about something scary that has happened to earth's cycles in the last few hundred years. That something scary is pollution and waste.

Pollution that is pumped into the air, dumped into the water, or sprayed onto or leaked into the land can harm parts of the cycle. When that happens, the cycle stops working properly, much as your bicycle tires stop working when you run over a nail. Because all earth's cycles work together, harm done to any one cycle harms the other cycles, too.

Water gives plants and animals the moisture they need, and air carries the gases they need.

Plants provide food and oxygen for animals.

Animals eat the food and breathe in the oxygen. When they die, their bodies rot and enrich the soil.

The sun gives plants energy to grow.

The soil provides food for new plants to grow.

Waste also damages earth's cycles. The world can be seen as a giant cafeteria for people, offering us the choice of all the forests, oceans, vegetables, animals and minerals it contains. But people are greedy customers—we've taken more than we need and the natural world is suffering as a result.

Pollution and waste harm earth's cycles. Fortunately, unlike your bicycle tires, the Earth can sometimes repair itself. But that can happen only when the pollution is removed and the wasting of resources is stopped. And that's where you come in.

Read this book to find out how Earth's cycles work, then follow the action trail (it looks like this) and see how you can help stop pollution and waste. Kids working together *can* make a difference!

What are ecosystems?

What has water weeds for lungs, river currents for veins, the sun's rays for food and rainwater for drink? No, it's not a monster from outer space—it's a wetland ecosystem.

Ecosystems are like giant creatures that eat, drink and breathe. They don't do this on their own—millions of different plants and animals help them do each thing.

Working together, living and non-living things form ecosystems. In this chapter, you'll find out how ecosystems operate.

Each part of an ecosystem—animals and plants and the air,

A wetland is an ecosystem. So is a lake, pond, river, or stream.

water and earth that sustain them—plays an important role. No part can survive without the others, just as your hand couldn't live if it was separated from the rest of your body.

All living things depend on one another. But, sadly, we're different from the other plants and animals that we share the world with. People are great at taking from the natural world; we're not so good at giving back to it.

We have to change the way we look at Earth's ecosystems. They're not just there to serve us. The animals, plants, the wilderness, the undisturbed beauty of the natural world have a value all their own—a value that we can't measure in our human currency of dollars and cents. And there are lots of things that we can do to show more respect for the planet we share. The next chapter shows how we people have altered the natural systems we depend on, and how we can help to restore their original balance.

A beach is an ecosystem. So is an underwater kelp forest and a coral reef.

A desert is an ecosystem. So is a forest.

The right stuff

What contains:
½ bathtub full of oxygen
50 glasses of water
125 mL (½ cup) of calcium
125 mL (½ cup) of sugar
1/10 thimble-full of salt
a tiny pinch of assorted elements such as
  nitrogen, phosphorus, potassium, sulphur,
  magnesium and iron
a mystery ingredient

Believe it or not, these are the basic ingredients that *you* are made of. Buying these ingredients in a store would cost less than $20, but of course no one can measure your value in dollars and cents. What makes you very special is the unique way these ingredients are put together.

Mix together eggs, butter, flour, sugar, milk and baking powder or soda in different ways, and you can make a pie crust, a cake or cookies. Combine the basic ingredients of a human being and you get hearts, muscles, blood, skin, hair or bones. Your aunt, your brother, your neighbour, even complete strangers are all composed of the same basic ingredients put together in different ways.

These important ingredients can be sorted into four groups:

- **soil**, including the things that make soil (rocks) and the things made *from* soil (such as bricks and glass)

- **water** and other liquids

- **air** and gases such as oxygen

- **the mystery ingredient**—energy (for more about this turn the page)

These are the basic building blocks of all living things, such as plants and animals, and of non-living things, such as rocks and lakes. So you have a lot in common with a worm or a rock. All three of you are made of the same building blocks—water, soil, air and energy. Because these basic building blocks are so important, we call them "resources."

93% OF YOU IS JUST WATER AND AIR!

The mystery ingredient

If you combined all the ingredients from the previous page together, it would take a miracle to mix them up into a human being or a worm. That miracle is energy—the power that enables living things to grow, move, eat and breathe.

You get energy from the food you eat. Where does the energy in food come from? It all begins with the sun. The sun helps plants turn water, carbon dioxide (a gas in the air) and nutrients into leaves through a process called "photosynthesis." Animals get energy by eating those leaves. You get energy by eating plants and animals. Without energy you couldn't survive, and neither could any of the other living things on Earth. Want to see what happens to living things when the sun is shut off?

Pick up a pail or other object that's been sitting on the grass for a week or so. What does the grass underneath look like? You'll probably find that it's beige or grey, if it's alive at all. Plants need the sun to produce the green stuff (called "chlorophyll") in their leaves. The chlorophyll traps the energy in sunlight, so it can be used to help plants grow. What else is different about these sun-starved plants?

Without the sun, animals would suffer, too. They might not turn beige or grey, but they'd get pretty hungry when all the plants that depend on the sun died. They'd suffocate, without the oxygen the plants release into the air, and they'd freeze, without the warmth of the sun's rays. Directly or indirectly, we all need the sun.

How are you like a dandelion?

As if being like a worm isn't bad enough, you're a lot like a dandelion as well. All living things have seven special qualities, and each of these qualities depends on energy.

1. *They make new beings, like themselves.*
2. *They breathe.*
3. *They eat.*
4. *They respond to changes in temperature, light and wetness.*
5. *They can move. (Even plants grow towards the light!)*
6. *They get rid of extra water. (They pee or release water droplets.)*
7. *They grow.*

CHASE THE SUN

Plants need sunlight to grow and produce food for us and other animals. You can see how important sunlight is to plants by trying this experiment.

YOU'LL NEED:
scissors
a cardboard box with a lid, big
   enough to hold the plant
a small indoor plant with a
   long stem
a window sill

1. Use the scissors to cut a hole big enough to put your fist through in one side of the box.

2. Put the plant in the box and close the lid. Put the box on the window sill.
3. Water the plant twice a week. After a few days, remove the plant from the box and look at its stem and leaves.
4. Your sun-starved plant twists and turns to find the sun. Try putting the plant into another box with a hole in a different place and see what happens.

AS PLANTS GROW, THEY ALSO PRODUCE OXYGEN, WHICH WE BREATHE.

15

Natural needs

How would you feel if you never ate another ice-cream cone or drank another soft drink? You might be unhappy, but you'd survive. You can live without soft drinks and ice-cream cones, but you can't live without air, water, soil and energy. These are the essential resources that all living things need. Take away just one of these resources and you'd be in trouble. Try it and see, using a plant as a guinea pig.

THIRSTY SEEDLINGS

YOU'LL NEED:
2 paper towels
a plastic bag
a package of radishes or beans seeds
a window sill
4 small plant pots filled with soil (yogurt containers with holes punched in the bottom work well)
a pen and labels
4 saucers

1. Fold two layers of paper towels in half and wet them thoroughly. Lay the paper towels out flat in a plastic bag and sprinkle the seeds on top of the wet paper towels.

2. Fold over the end of the bag to seal it and put the bag on a sunny window sill.

3. When 10 seeds have sprouted (produced a seedling with a couple of tiny leaves and a thin root), plant them two to a pot.

4. Label the pots Soaked, Wet, Moist and Dry. Then place one pot in each saucer.

5. Water the Soaked pot with 125 mL (½ cup) of water every day. Water the Wet pot with 125 mL (½ cup) of water twice a week. Water the Moist pot with 125 mL (½ cup) of water every week. Don't water the Dry pot at all.

Watch what happens to the plants. How much water keeps them healthiest? Can you have too much, as well as too little water?

Some plants can survive with less of one resource than other plants. Test to find how much water different kinds of plants need to be healthy by sprouting and watering other types of seeds. Try dried beans, lentils or herb seeds such as mustard, coriander, caraway, sesame, cardamom or poppy seeds.

Not a drop to drink

All living things need water. What happens when water is scarce? Meet two desert dwellers who have come up with ways to survive in the driest places on Earth.

Chilean cacti in the Atacama desert can go without water for 400 years! But when it rains,

HUNGRY PLANTS

All plants and animals need food to grow and survive. Each creature gets its nutritional needs from different amounts and types of food. Try this to find out which nutrients one variety of plant requires.

YOU'LL NEED:
8 seeds from a package of
  bean or radish seeds
2 paper towels
a plastic bag
4 small plant pots (yogurt
  containers with holes
  punched in the bottom
  work well)
sand (enough to fill two
  plant pots)
soil (enough to fill two
  plant pots)
compost (enough to half fill
  a plant pot. If you don't
  have compost, grind up
  carrot peels, grass or
  leaves.)
a large cake pan

1. Sprout your seeds by following steps 1–2 on page 16.

2. Fill two plant pots with sand. Mix the soil and compost together, and fill the other two pots with this mixture. Place all four pots in the cake pan and add water to a depth of 1 cm (1/3 inch).

3. Plant two seedlings in each pot. Place the containers in a sunny spot and water them every few days.

The sandy pots contain almost no nutrients (food). The pots with a mixture of soil and compost are full of nutrients. Which ones do your seedlings prefer?

the cacti really soak it up. Some become 90% water and use their built-in water supply to survive the next dry spell.

Kangaroo rats never need to drink at all! They live in the desert and dry plains of western North America, and they get all the water they need from the seeds they eat.

Homes and habitats: your biological neighbourhood

Where do you find the resources you need to survive? You go to the fridge (or the store) for your food and to the tap for water; the air you breathe comes from all over your neighbourhood. The location of these resources defines your "habitat."

Your habitat is more than just your house or apartment. You eat in the kitchen, you play outside in the park, yard or street, and you learn in your classroom. Where else does your life take you? To shops, the library, your friends' homes? All these places make up your habitat—your biological neighbourhood.

All animals and plants have their own habitats. An animal's habitat is the place where it breeds, raises its young, finds food, water and shelter. Habitats can be huge or they can be tiny. Salmon need whole ocean and river systems for their habitat, but the mistletoe plant spends its whole life on one oak or sycamore tree. This one tree provides everything the mistletoe needs in life—it is the mistletoe's entire habitat.

BUILD A HABITAT FOR A BUG

You can build a habitat for an insect or other small creature, and add everything it will need to survive.

1. Choose an ant, or some other animal that lives in the soil.
2. Watch your creature in the wild, and read up on its habits and habitat in a bug book.
3. Line a cardboard box with tin foil, and add earth, mud, grass, twigs, dead leaves and anything else that your creature needs to survive.

You've built a habitat for your bug. How does your bug's habitat compare with your own?

Habitats high and low

Our habitat covers a large area, close to the ground. Some animals, like the cicada, have habitats that extend up into the trees and down underground.

You might never have seen a cicada, but you've probably heard their buzzing song on a hot summer day. Cicadas inject their eggs, which are the size of very small rice grains, into twigs high up in trees about 500 at a time. When the eggs hatch in July, the nymphs (baby cicadas) drop to the ground. The tiny blind creatures quickly burrow into the earth to avoid being burned by the heat of the mid-summer sun. Each nymph builds a tiny cave for itself several centimetres (inches) underground. Sucking sap from tree roots for food, the nymphs live underground for 17 years. Finally, they emerge from their dens, crawl up their tree, make their distinctive buzz, mate and lay eggs. In their whole lifetime, they travel only a few metres (yards) up in a tree and about 10 cm (4 inches) underground. They rarely leave the tree where they were born.

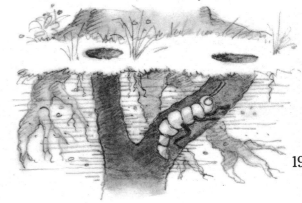

19

Biological business

Who shares your habitat with you? Your family, friends, neighbours and schoolmates? How about other animals and plants? Do you have any pets or unwelcome house-mates, such as fleas, flies or ants? Do you have houseplants, herbs or vegetables in your home or garden? All these living things share your habitat. They form your local community—a community of living things that live together and depend on each other for their survival. Who do you depend on? Does anyone or anything depend on you?

Your community is a different place because you live in it. You play an important role: you water the garden, feed the cat, swat flies. Some things you do may also be harmful. For example, the garbage you throw out clutters and pollutes the land and water where it ends up. Other creatures in your community play roles that affect you: your parents buy the food you eat, mosquitoes sting you and use your blood to feed their young, and trees shade you from the hot summer sun. The changes that you make in the world and that the world makes in your life define your biological role. Ecologists call this your "niche."

What (or who) do you eat, meet, kill, grow, feed or use to build with, wear, sit on or shelter beneath? Trying to live on your own would be a bit like playing a baseball game by yourself. You need other players to have a good game, and you need lots of other creatures to live.

Imagine life without earthworms, for instance. How would it change—apart from the shortage of fishing bait? There'd probably be a lot fewer plants, for a start. Why? Earthworms crumble up the soil so plant roots can penetrate it. They recycle dead plants by eating them and spewing them out in a form that acts as plant fertilizer. They mix the earth up and create gaps for air and water that the plants need. All in all, the insignificant earthworm does the rest of the community a lot of good.

Worms at work

Here's how you can watch a worm recycling some leaves into fertilizer and creating tunnels to let air and water into the soil.

YOU'LL NEED:
a glass jar at least 25 cm (10 inches) deep
enough garden soil to almost fill the jar
6 worms
decaying leaves and lettuce
a piece of black paper big enough to fit around the jar
tape

1. Fill the jar loosely with soil. Leave about 5 cm (2 inches) of space at the top. Dampen the soil with water. It should not be soaked, just moist.
2. Put the worms on top of the soil and cover them with a layer of decaying leaves and lettuce.
3. Wrap the jar with black paper and tape it in place. This will fool the worms into thinking they're underground.
4. Put the jar in a cool dark place. Dampen the soil every day, but do not otherwise touch the jar for at least a week. Worms do not like to be disturbed. As the leaves and lettuce disappear add some more. These are the worms' food.

5. After 10 days, unwrap the black paper and look at what the worms have been doing. Do you see tunnels that the worms have dug? These allow water and air to penetrate the soil. The worms also eat the leaves and turn them into fertilizer.
6. When you are finished observing your worms, return them to the garden.

From dirt to dessert and back again

What does it take to make an oatmeal chocolate-chip cookie? If you really plan ahead, all you need are some seeds and soil, water, sunshine and air. Sounds more like a mud pie than a cookie you'd want to eat, but that's just how your cookie started out. In fact, those same four basic resources (soil, water, sunshine and air) fuel all life, creating plants that are then eaten and turned into animals, whose bodies and wastes turn into more raw resources. In nature, nothing is wasted. Materials are recycled from creature to creature, and resource to resource. Here's how.

Your cookie started out as soil, water and air. Oat, cocoa and wheat seeds were planted in the dirt, were watered by the rain and germinated in the sun. With a little help from farmers and food factories, the oat, cocoa and wheat plants were made into oatmeal, chocolate and flour. Chickens and cows turned other plants into eggs, milk and butter. Before you knew it, you had a cookie that'd make your mouth water.

So, go ahead. Eat your cookie. YUM! After you've digested it, you turn it into urine and solid waste that goes into the sewers. These wastes are purified and, eventually, become water and soil for new plants. Soon you're on your way to another cookie!

Plants, animals and sunshine turn soil, water and air into the food we eat and then into soil, water and air again. We recycle resources. All living things carry out this useful job.

When wild animals recycle, they make new food for other plants and animals. For instance, grizzly bears eat and drink. Most of this food and water gets turned into more grizzly bear. The rest ends up as urine and solid wastes, which other animals use as food and drink. For example, an earthworm might eat the bear scats (solid waste) and recycle them into rich soil, which acts as plant food. Around and around resources go.

22

THE SELF-WATERING PLANT

Want to see some water get recycled? Here's how to make a miniature version of a water cycle—and get a plant to water itself!

YOU'LL NEED:
a small potted plant
a plastic bag that will completely cover the plant and a twist tie

1. Water the plant as you normally would.
2. Put the plant in the plastic bag and seal the bag shut with the twist tie.
3. Leave the plant where it usually sits and watch what happens over the next few days.

The plant absorbs water from the soil. The water travels up into the leaves, where it is "breathed out" by the plant. The water vapour can't escape. It sticks to the plastic bag and eventually falls back onto the plant or the soil. It has been recycled!

23

Breeding like rabbits

Did you know that a mother cottontail rabbit can have as many as 30 babies in her life? Imagine if human families were that big! So why isn't the world over-run with rabbits? Because rabbits live dangerously. Only a very small number out of 30 baby rabbits will live long enough to have their own babies. Diseases, starvation and other animals kill off most of the rabbits. So there are usually only enough young rabbits to take the place of their parents.

Thousands of years ago, human families were a bit like rabbits. Only enough children survived to replace their parents. But nowadays, the number of people is growing fast. We have controlled most dangerous diseases by vaccination, better nutrition and, most importantly, by cleaner living habits.

In Zimbabwe today, parents replace themselves four times over; they have an average of 8.2 children. Overall, in the developing world (not counting China), the average number of kids per family is five. The world average is 3.7.

The good news is that most of us get to live long, healthy lives. The bad news is that while we enjoy ourselves, a lot of the natural world suffers.

As our families grow, so does our use of natural resources. North Americans are especially greedy when it comes to resources. In total, an average of at least 40 tonnes (tons) of food and 26 million tonnes (tons) of water are consumed by just one North American in a lifetime. And food isn't the only resource we gobble up. We use the equivalent of nearly 7000 kg (15 500 pounds) of coal, much more than we actually need to survive.

Luckily, most people from other countries aren't as wasteful as we are. In Madagascar, people use less than 2 million tonnes (tons) of water—about one-twentieth of what we use—in their lifetimes. Asian energy use is only one-tenth of ours.

Recently, the human population explosion has begun to slow down. Experts guess that by about 2050, the number of people in the world will stop increasing. But by then, there will be 10 billion of us, twice as many people as there were in 1988. That's a lot of bodies to feed, fuel, clothe and shelter. Where will all the resources they need come from?

USE IT UP!

How many resources does a family use over three generations? Get some marbles and head for a sandbox to find out. A marble represents one person. Each person needs energy, food, water and shelter to survive. These resources are represented by a handful of sand.

1. Start with the two great-grandparents. Set out one marble for each of them. Now pile up a handful of sand for each.

2. The great-grandparents have four children. Add a marble and a handful of sand for each child. These are the grandparents.

3. The grandparents marry and each has four children. Add a marble and a handful of sand for each child. These are the parents.

4. The parents all marry and each has four children (still more marbles and sand).

What are you left with? A small number of marbles (people) and a large mountain of sand (resources used).

In less than a hundred years, the family has grown from two to several dozen. All those great-grandchildren will need a lot more food, water, space and shelter than the great-grandparents needed. If you kept going, you'd run out of sand. That is what has happened to us! We human beings have multiplied too quickly, and we consume so many resources that we're running out of them.

A TALE OF TWO FAMILIES

Here's what a family tree for a stone-age family that lived 100 000 years ago looks like compared with a modern-day family from Zimbabwe. In just three generations, there are more than ten times the number of children in the Zimbabwean family than in the stone-age family.

Stone-age family

Modern-day Zimbabwe family

Disturbing actions

The natural world is always changing. For example, every year beavers dam streams and drastically alter the environment for the plants and animals that live in or near the water. The dams, which can be as tall as a house, make the stream water back up and flood the nearby area. Trees drown and become dead wood. Submerged plants rot and new water plants move in. Black flies, midges, mayflies and caddisflies cannot survive in the new pond environment. They die and are replaced by dragonflies, tubificid worms and clams. The new beaver pond provides the perfect habitat for otter, mink, muskrat, moose and ducks. The changes are gradual and natural. Plants and animals either adapt to the new conditions or are replaced by other species.

Sudden changes can be much more damaging. When we humans build huge hydro-electric power dams, whole communities of living things can be wiped out. Huge lakes covering dozens of square kilometres (miles) are created. The dams release water from time to time to create electricity. These continual changes to the water level make it difficult for healthy communities of plants and animals to survive. The ecosystem can be damaged forever.

The bigger and more complex the ecosystem, the less it will be affected. Why? One reason is that it's difficult to disturb all the inhabitants of a very big community. For example, it would take a long time to pollute Lake Superior because it is so big and contains so much water. Scientists calculate it takes 191 years for the water in Lake Superior to be replaced as new water flows in and old water flows out. Things happen slowly in such a big lake. But the water in smaller Lake Erie is completely replaced in just six years. It's at much greater risk.

When communities are disturbed, they *can* return to a healthy state—if the cause of the disturbance is removed. But if too much is destroyed the ecosystem is changed forever.

DISTURB YOUR LAWN

Find out what happens when you suddenly change a small patch of nature.

YOU'LL NEED:
some lawn
tent pegs or pencils or broken pens
a lawnmower or some shears
a trowel or a large spoon

1. Measure out three patches of lawn, each about the size of a pizza box. Mark the corners of each patch with pegs, pens or pencils. (One of your patches of lawn may get damaged for a while. Check with your parents about an out-of-the-way location for this patch.)

2. Mow one patch using your shears or a lawnmower, leave another one to grow, and remove all the plants from the third, using your trowel or spoon to dig the roots out.
3. Check out the three patches after a week to see which ones look like they used to.

Usually, the more drastic the disturbance, the more difficult it is for the community to return to its former state. Did you find this with your harassed lawn?

Go back and take another look at your lawn patches after a month. Your three patches will likely become more and more similar the longer you leave them. Animals and plants and even entire communities can recover from disruption, if they haven't been treated too badly. Even the hungry, thirsty plants from pages 16 and 17 can be revived if they're given what they need soon enough.

A tale of two forests

This forest on the left has been clear-cut—all the trees have been completely cleared out. Large bulldozer-like machines drove into the area, the trees were chopped down with chainsaws, then huge machines cut off their branches. The logs were carted off to a sawmill or pulp mill to be turned into boards or paper.

When a forest is clear-cut, many kinds of plants and animals lose their homes; they must leave or die. Even the soil starts to disappear. Rainwater flushes through the logged area, and without the tree roots to hold down the soil, the water can wash the earth away. Loose dirt floods into nearby streams, filling them with muck. The soil left in the area where the forest used to be is now much poorer; the rain has washed away much of the valuable plant food that it normally contains.

WHEN TREES ARE REMOVED, THE LAND MAY BECOME A DESERT. SIX MILLION HECTARES (14.5 MILLION ACRES) OF DESERT ARE CREATED THIS WAY EVERY YEAR.

WITHOUT TREES, THERE'S NOTHING TO HOLD THE SOIL IN PLACE. IT BLOWS OR WASHES AWAY, LEAVING LITTLE GROWING SOIL FOR NEW PLANTS.

It doesn't have to be that way. There are ways to preserve the forests that are so valuable to us. Loggers could cut only some of the trees in an area, using a method called selective cutting. This way, they would visit each patch of forest every so often and cut down and remove only the oldest trees, which are ready to be harvested. Or they could clear-cut only small parts of the forest so that seeds from nearby trees could help with reforestation. These methods of logging take longer and cost much more than clear-cutting, but they allow the forest to regenerate naturally and don't disturb the animals and plants that live in it.

Careful logging is a wise use of resources. Nature is harvested in ways that let it recover from our visits. This is called "sustainable development." It's better for the environment, and it's better for us, too, because it guarantees that there will always be resources for people in the future.

IF PEOPLE USED LESS WOOD AND PAPER OR RECYCLED WHAT THEY DID USE, MANY TREES COULD BE SAVED.

29

Now you see it

Have you ever seen a North American paddlefish? Probably not, except in books. The paddlefish is in danger—there are very few left in the whole world. We've nearly killed them off by spoiling their habitats and harvesting their eggs for caviar. In Canada, paddlefish are more than endangered —they are "extirpated." This means there are none left at all.

In the past 200 years, more than 50 kinds of birds, 75 kinds of animals and probably hundreds of kinds of insects have become extinct. Extinctions are becoming more and more common, and it's now estimated that one organism bites the dust each year. (About 1 000 kinds of mammals and birds and one out of every ten flowering plants are threatened with extinction.) By the year 2000, one of every five species on Earth will have become extinct.

People are largely to blame. We've fouled animals' nests, poisoned their habitats with pollution, and introduced foreign species to their homes, which eat their food or kill them off. In some cases, we've completely destroyed their living spaces by chopping down, flooding, or burning their homes. A lot of this destruction has happened in the tropical rainforests.

Only 6% of the Earth's land area is rainforest, but these rich forests contain half of all the kinds of life in the world. They provide a great deal of the oxygen that we need to breathe, and rainforest plants are important natural sources for new medicines. Every day, more than 40 400 hectares (100 000 acres) of rainforest are destroyed! That's about 40 hectares (100 acres) every minute! Rainforests are cut down to make space for everything from mines to cattle ranches.

The trouble is that, in the tropics, when the rainforest is logged and the soil left bare, the earth may turn into a hard, crusty substance called laterite. It's so tough and solid that it has been cut into bricks and used to build temples in Asia. Plants have a tough time growing in this construction-grade soil. As a result, logged rainforest can quickly become useless for any sort of farming. And the animals and plants that used to live in the rainforest can never return.

Paddlefish

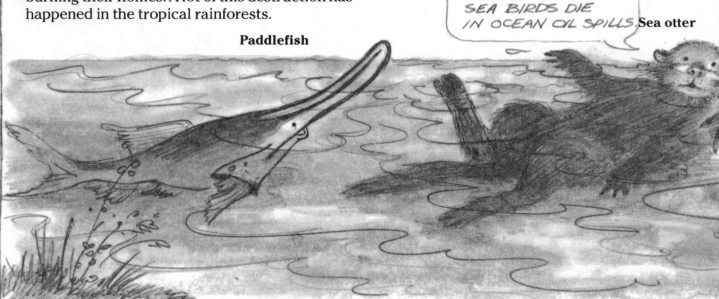

SEA OTTERS AND SEA BIRDS DIE IN OCEAN OIL SPILLS **Sea otter**

THE EGGS OF SOME BIRDS ARE DAMAGED WHEN THE PARENT BIRDS EAT INSECTS SPRAYED WITH FARM CHEMICALS.

Hawk

When plants and animals disappear, we lose something. If a rainforest is cut down, we may lose plants that give us life-saving medicines. Draining a marsh or polluting a lake will wipe out other species that add to the beauty of our world. Diversity is important; we need lots of different plants and animals.

Does your neighbourhood have lots of plant and animal diversity? Check it out by counting the number of different kinds of bugs, plants or birds you find in a local playground. Now do the same thing in a wilder place, such as a wooded park. Where do you find more plant and animal species?

Kit fox

CITIES AND FARMS GOBBLE UP WILDERNESS... AND ANIMAL HOMES.

Helping nature to live with us

There are things that can be done to make life easier for the plants and animals that we disturb with our building projects. One is to help them dodge cars! When we carve roads through forests or fields, the animals that live there are cut off from part of their home by the speeding vehicles. So people have made special routes for them to carry them safely past the cars.

• In England, more than 150 000 toads are squashed each year by cars on highways. So special toad tunnels have been built under the roads. These tunnels can accommodate 200 toads per hour. Badgers and hedgehogs can use the toad tunnels, too.

• In Banff National Park, enormous underpasses have been built so that large animals such as deer and elk can cross the Trans-Canada Highway safely.

CHAPTER 2:
AIR

Take a deep breath. If you think you're breathing in nothing but air, guess again. The air you've just sucked in is a complicated combination of chemicals, plant seeds, tiny organisms and bits of animal, plant and mineral matter.

Air is an important chemical highway: stuff that we need, such as oxygen, is brought to us in the air, and stuff we'd be better off without, such as factory fumes, smoke and car exhaust, comes right along with it.

The air carries pollution across continents and oceans. For better or worse, we have to share our air with the rest of the world.

2. **Water falls from clouds that form in the air. The bits of soil fall with it onto the land and plants.**

3. **Plants absorb water, the nutrients it carries and air. As they grow, they release oxygen that animals need to breath.**

1. The air absorbs water from lakes, rivers and oceans. It also carries bits of soil from the land.

5. The air constantly moves around the earth.

4. People and animals breathe in the oxygen.

Because air travels, it's very tough to keep it clean. Air pollution may be banned in one place, but polluted air can blow in from another. Cleaning up the air is a tough job, but it's not impossible. Read on to find out how you can help.

Collect indoor pollution

Did you know that you are breathing air that the dinosaurs breathed millions of years ago? Air is recycled. It goes in and out of animals' and people's lungs, generation after generation, century after century. You need the oxygen that air contains. But only one-fifth of the air you breathe is oxygen. The rest is mainly nitrogen, with bits of carbon dioxide and water vapour thrown in. If that were all, your lungs would be happy. But it isn't. By driving cars, burning fires and running factories, people have added lots of invisible substances—called pollution—to the air. And what you don't see *can* hurt you.

Pollution particles are often so tiny that your nose hairs and other body filters can't catch them. You breathe them right into your lungs, where they can do real damage.

Pollution doesn't stop at your front door. It comes right into your house and makes itself at home. You may even add more pollution when you paint or buy new furniture or carpets that release fumes.

What can you do to get rid of indoor pollution? Cleaning up your room might not be your favourite pastime, but when you dust, vacuum and wash the floors frequently, you remove all the pollution that's attached to the dust in your room (and lots of it is). If you live in a house that is heated and cooled with air ducts, ask your parents to clean the ducts and change the air filters often.

Clean up, and you can start to control indoor pollution. Unfortunately, you can't control the *source* of the pollution. Governments try to stop pollution by passing laws to limit how much pollution people and factories are allowed to release into the air. But it isn't easy. First, they have to figure out how much pollution human beings can put up with (and not everyone agrees about this). Then they have to write tough laws telling companies to clean up. Finally, they have to take polluters to court.

Often there are big problems keeping pollution under control, but it's not hopeless. You can make a difference, by encouraging your politicians to get tough with polluters, by complaining if you see, smell or taste pollution, and by getting together with your friends to carry out environmental work. To find out more about how you can help, see page 88.

POLLUTION BUSTERS

You can trap some of the pollution in your house with the help of a vacuum cleaner.

YOU'LL NEED:
a saucer
a piece of fine white cloth about the size of
*   a hand towel (an old bed sheet works*
*   well)*
a pencil
scissors
a vacuum cleaner with a hose attachment
2 large rubber bands

1. Place the saucer upside down on the cloth and draw around it with a pencil. You'll need to draw four of these saucer-sized circles on your piece of cloth. Cut out the four circles.
2. Wrap one circle over the nozzle of the vacuum cleaner. Secure the cloth with the elastic bands. The cloth should cover the hole and be held tight on all sides like a drum-skin.
3. Turn the suction on and vacuum the floor for a couple of minutes.
4. Turn off the vacuum cleaner and remove the piece of cloth. Is it still white? The dirt on it came from the air, from your clothes, from your feet, from pets' fur or from cigarettes and landed on the floor. The same stuff gets swirled up into the air and goes into your mouth, nose and lungs every day.
5. Try Steps 1–4 again, vacuuming in different spots—behind and underneath the furniture, in the kitchen and basement and attic. Check and change your cloth after each area is vacuumed. Where is the floor the dirtiest? The cleanest?

Travelling toxics

Most kinds of pollution that travel by air are tasteless, odourless and invisible. But here's a way to see some of the bigger bits that pollution often sticks to.

YOU'LL NEED:
a clean bucket about half full of clean water
a finely woven nylon stocking or a coffee filter to act as a pollution filter
tweezers
a sheet of clean white paper
a magnifying glass, if you have one

1. Set the bucket of water outside in a spot where it won't be disturbed. It should be out in the open.
2. Leave it for a week.
3. To get a good look at the stuff on the bottom of your bucket, pour the water through your pollution filter. Then turn the filter inside out and use your tweezers to spread the dirt out on a piece of white paper. What did you trap?

Long-distance landings

Most kinds of pollution are too small to see. But one type of airborne pollution is easy to spot. Balloons filled with air or helium gas can travel thousands of kilometres (miles). That would be pretty neat, if balloons didn't cause harm when they landed. But some do. Balloons can end up in the stomachs of dolphins, turtles, birds and whales, killing them slowly and painfully. So, next birthday party, make sure your balloons end up in the garbage, not in an animal's stomach!

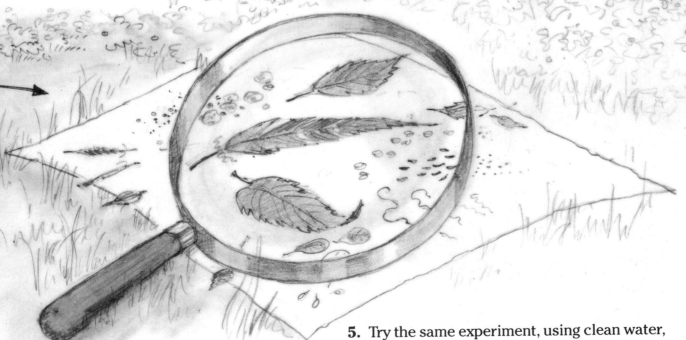

4. If you have a magnifying glass, look through it and see what you can identify.

5. Try the same experiment, using clean water, but put the bucket in a different spot. Try putting it near a street, then away from the street. Do you catch different things? More things or fewer?

Smoky air

Most air pollution comes from factories, cars and power plants. But some pollution is produced by individual people. Drivers, barbecuers, bonfire makers and smokers pollute. The smoke they produce affects everyone who works, plays and lives near them.

What does cigarette smoke do to people? Try it on a radish and find out.

YOU'LL NEED:
2 paper towels
a plastic bag
a package of radish seeds
6 small plant pots filled with soil (yogurt containers with holes punched in the bottom work well)
soil
cigarettes (explain to a grown-up what you want them for and ask them to buy some for you)
6 glass jars large enough to fit over the plant pots

1. Sprout and plant radish plants by following Steps 1 to 3 on page 16.
2. Put your pots in a sunny location and water them regularly.
3. Every evening, "plant" a burning cigarette in two of the pots. Cover the pots with a glass jar all night. Remove the jar covers each morning.

FACTORIES POLLUTE.

CARS POLLUTE

BARBECUES POLLUTE

4. Every evening, plant *two* burning cigarettes each in two other pots and cover them overnight. Remove the jar covers each morning.

5. Leave the seeds in the other two containers alone—no cigarettes, no covers, just regular water.

6. After two weeks of this treatment, do you see any difference in your plants?

Radishes don't smoke, but lots of people do. To find out how smokers are affected by their habit, ask a smoker you know to hold a white or light-coloured Kleenex over his or her mouth and exhale (breathe out) into it three times. What does the Kleenex look like afterwards? Then ask the smoker to blow three mouthfuls of smoke that hasn't been inhaled (breathed in) through another Kleenex. Notice how much dirtier the second Kleenex is? All that extra gunge usually ends up in the smoker's lungs.

IN 1965, OVER 40% OF ALL AMERICANS SMOKED.

TODAY, ONLY ABOUT 29% OF PEOPLE SMOKE.

MANY SMOKERS ARE QUITTING AND MANY KIDS ARE REFUSING TO START.

Migrating muck

Did you know that chemicals let loose in your neighbourhood could end up in Antarctica or Greenland? For example, lead spit out by cars using leaded fuel can travel thousands of kilometres (miles) and land in remote corners of the world. And polluted smoke pouring out of factory smokestacks can be sent high up into the air, where it can be blown halfway across the world. Try this experiment and see how.

YOU'LL NEED:
an empty toilet-paper roll and an empty paper-towel roll
aluminum foil
a large dark-coloured paper or cloth at least 1 m (1 yard) wide
    and 2 m (2 yards) long
a few spoonfuls of flour

1. Cover one end of the toilet-paper roll with aluminum foil. Do the same with the paper-towel roll.
2. Spread out the dark paper or cloth on the floor.
3. Place the two rolls, foil-covered ends up, near the shorter edge of the dark sheet.
4. Put half a spoonful of flour onto the top of the shorter tube.
5. Hold the bottom of the tube on the floor and blow across the top of it. See how far you can make the flour travel.
6. Repeat Steps 4 and 5 with the taller tube. How far does the flour travel this time?

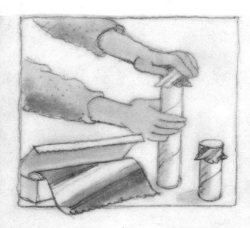

Factories discovered the same thing you did: the taller the tube (smokestack), the farther the pollution travelled. This kept the factories' close neighbours happy, but unfortunately the pollution wasn't really reduced. It was just pumped higher into the air and blown farther away. By the time the pollution landed, it was far from the factory that produced it.

Today, many factories spew out less pollution than they used to. The harmful wastes are trapped before they go up the smokestack. Not only does this help keep the air clean, it often saves the factory money, too, because some wastes can be sold or re-used.

Smoke + Fog = Smog

Among the worst of the travelling toxics is a chemical called sulphur dioxide. It's one of the causes of acid rain and also helps create smog—a nasty mixture of smoke and fog that sometimes settles over cities. In 1948, the smog in Donara, Pennsylvania, was so bad that people in that town couldn't even see as far away as their hands! An even more serious smog settled over London, England, in 1952 and killed 4 000 people. The air over Mexico City is so filthy that birds flying through it sometimes drop out of the sky dead! Such killer smogs are rare, but even smaller amounts of smog can still cause harm, damaging buildings, crops and trees, and risking the health of old people and babies.

41

Coping with cars

Get up and look out the nearest window. If you can see a street, chances are you'll see cars on it. There are about 400 million cars in use today around the world. They're the main source of pollution in the cities. They release:

• Carbon monoxide, a chemical that can hurt humans if they breathe too much of it. It makes people breathless, slower and duller by cutting down on the amount of oxygen that gets into the blood. Too much of it can make it difficult for people to breathe or can even kill them.

• Nitrogen and sulphur oxides, leading causes of acid rain and acid snow, of city smog and of suffering in people with lung problems.

• Tiny bits of other materials, such as lead (now banned in many countries), which can also cause harm to people and the environment in general.

In Florence, Italy, cars aren't allowed in the centre of the city from 7:30 a.m. till 6:30 p.m.

Bikes have largely taken over from cars in Kasukabe, Japan. There are so many bikes that they have to be hoisted by cranes into special parking garages.

In other cities, cars are so slow it's a wonder anyone bothers with them at all. Bicyclists in London, England, can beat cars, which average only 13 km (8 miles) per hour.

Automobile pollution is so bad in some cities that cars have been banned! How can you reduce car pollution? Try using muscle power instead of horse power: walk, jog, cycle or skate to your school, sports or other activities.

If it's too far or too dangerous to travel these ways, try the bus, streetcar or subway. Public transit carries more people than private cars can and produces a lot less pollution. If you have to use a car, make sure it's full: car-pool with friends to school, sports or other events.

Hidden pollution

You can smell (and often see) the pollution that gushes out of car tailpipes. So you know it's there. But there's hidden pollution in cars, too. Each year, nearly 100 million kilograms (200 million pounds) of plastics are used to make cars. To make each kilogram (pound) of plastic, two kilograms (pounds) of hazardous waste are created. That's nearly 200 million kilograms (400 million pounds) of waste to get rid of—a huge problem for all of us.

Car pollution doesn't stop when a car dies and gets hauled off to a junkyard. Rusting car bodies clutter the landscape, and piles of used tires can turn into an environmental disaster. On February 12, 1990, 13 million used tires caught fire and burned for 17 days at a tire depot near Hagersville, Ontario. Oil and dangerous compounds melted from the tires and seeped into the land and groundwater below, causing serious contamination. A tire plant in Ohio dumped so much hazardous waste into the Cuyahoga River near Cleveland that the water caught fire in 1969. Happily, two decades after that fire, the river is recovering.

Put yourself in the driver's seat

A game for 2 or 3 players

Can you make it from Home to School first, without picking up any pollution points? Get some friends together and try this board game.

YOU'LL NEED:
three different colours of buttons to serve as markers
a single die (not a pair of dice)
a piece of paper and a pencil

Rules:
1. You can drive either a CAR, a BUS or a BICYCLE from Home to School. Decide which you want to drive and choose a button marker as your vehicle.
2. Each player rolls the die. Highest roll goes first.
3. CARS and BUSES move twice the number shown on the dice. BICYCLES move the number shown. (After all, cars and buses travel faster than bicycles.)
4. When you land on a square that has instructions for your vehicle, follow the instructions. If the instructions tell you to collect pollution points, write down the number of pollution points on the piece of paper.
5. All players must get to School before the game is over. Keep track of who comes first, second and third.

How to score:
1. First to finish gets 10 points
   Second to finish gets 8 points
   Third to finish gets 5 points
2. Subtract pollution points from the points you earned for finishing the game. So, for example, if you finished second but collected 3 pollution points, you would finish the game with 8−3=5 points.

Play the game several times and add up the number of points for each vehicle. Which vehicle has the highest score?

HOME

BUS stop. Miss a turn.

On BIKE? Take a shortcut.

No CARS or BUSES allowed.

No CARS or BUSES allowed.

Flat tire. CAR & BUS collect 2 pollution points.

Detour! CAR & BUS follow arrow and collect 3 pollution points.

Dirty BUS exhaust. Collect 2 pollution points.

Icy road. CAR, BUS & BIKE go back 4 squares.

CAR breaks down. Go back to HOME.

Gas Station. CAR & BUS miss a turn & collect 3 pollution points.

Road repairs. CAR & BU

Chain breaks on BIKE. Miss a turn.

Parade! CAR & BUS follow detour.

CAR breaks down. Go to Gas Station.

Traffic jam! CAR & BUS collect 4 pollution points.

CAR out of gas. Go to Gas Station.

CAR stalls. Collect 2 pollution points.

1 L (¼ U.S. gallon) of gas can move one passenger 10.5 km (6.5 miles) in a car, 42 km (26 miles) in a bus, or the equivalent of 425 km (265 miles) on a bicycle.

BUS stop. Miss a turn.

SCHOOL

Detour. CAR follow arrow.

BIKE has flat tire. Miss a turn.

BUS transfer. Miss a turn.

CAR out of gas. Go to Gas Station.

Accident! CAR & BUS miss a turn.

Gas prices up. CAR & BUS go to Gas Station.

BUS stop. Miss a turn.

CAR leaves oil spill. Collect 4 pollution points.

On BIKE? Speed up and go ahead 3.

Special BIKE lane. Go ahead 2.

iss a turn.

Detour. CAR & BUS go to Gas Station.

Traffic jam! CAR & BUS collect 3 pollution points.

45

Protecting the earth's sunscreen

Have you ever used sunscreen? It's the goop that people smear on at the beach to protect them from getting a sunburn. It works by blocking out the sun's dangerous rays. You may be surprised to learn that Earth has a protective layer of sunscreen on all the time. It's called the ozone layer, and it's a covering of sun-blocker about 20 to 50 km (12 to 31 miles) up in the sky.

Without Earth's sunscreen, we'd be in big trouble. Gradually, the Earth would get sunbaked. At first, you might enjoy getting a quick tan. But eventually, the extra sunlight would cause big problems. Farm crops would be damaged by the radiation and would produce less food. People would be plagued with skin cancer and eye diseases, caused by the increase in ultraviolet rays.

Without its sunscreen, the ozone layer, the Earth would be "sunburned." Unfortunately, the ozone layer is in danger. Little by little, we are destroying it. There's already a big hole in the ozone layer over Antarctica. We destroy the ozone layer by producing chemicals that eat it up.

Chemicals called CFCs (short for chlorofluorocarbons) eat ozone. You can find CFCs in some spray cans, all fridges and air conditioners and in some of the styrofoam packages that food is packed in. We use a lot of these ozone eaters. By littering the Earth with ozone eaters, we are damaging the Earth's sunscreen.

What can you do to help? Cut down and recycle. Grab a pencil and paper and list all the things in your house that eat ozone. Could you do without any of these things? For example, instead of using styrofoam picnic plates, could you use ceramic ones? If you must use disposable, try paper plates or plastic ones. Better still, try to persuade companies that use ozone-eaters to use less. Don't buy their products or write them letters urging them to change their wasteful ways.

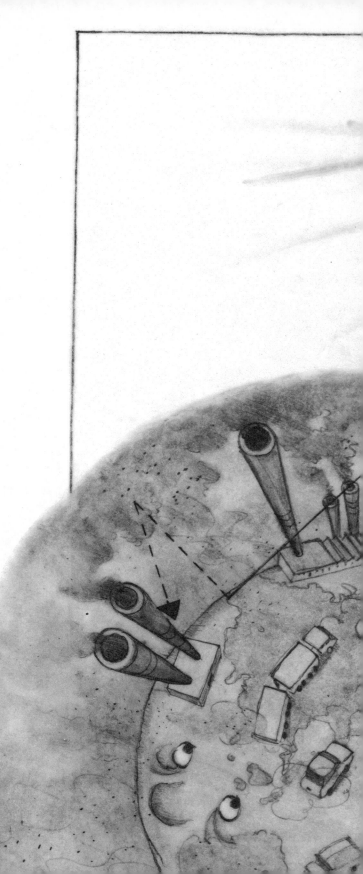

Bounce back

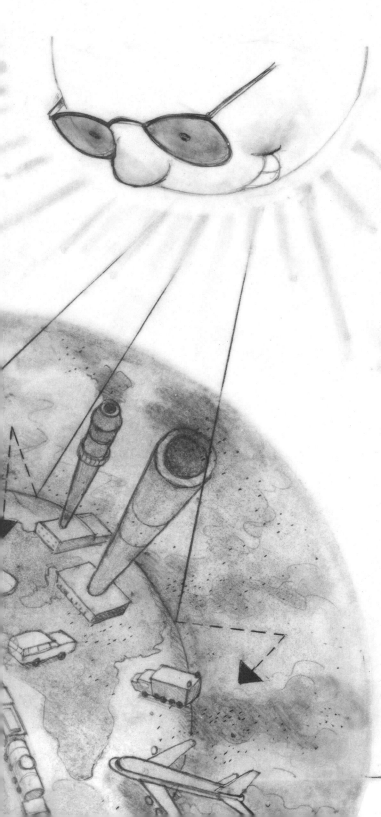

After rays from the sun hit the Earth, they bounce back towards the sun. Or at least they should. But lately, we've been dirtying up the atmosphere with exhaust fumes from cars, furnaces and power plants. These fumes contain a chemical called carbon dioxide, which traps the sun's rays and doesn't allow them to bounce back. The trapped rays heat up the Earth like a tomato in a greenhouse. That's why scientists call this warming the "greenhouse effect."

At first, the idea of extra heat might appeal to you. You might like to pack away the snow shovel forever and get out your summer clothes. But eventually, the warmth could build up until the whole Earth begins to dry up. Ponds and streams might disappear, and clean water could get even more scarce. Glaciers and polar caps might melt, expanding the oceans and causing floods of seaside areas.

To let the heat out, we need to cut down on burning coal, oil and gas in cars, furnaces and power plants. That means using less energy—to travel places, to heat and light our buildings and to power factories. There are some ways that you can help:

• Plant a tree: 2 kg (4.5 pounds) of carbon dioxide is used by every kilogram (2.2 pounds) of live tree.

• Use foot- and pedal-power instead of horsepower. For more on this, see page 43.

• Turn down the heat in winter and wear a sweater.

• Use less hot water by taking a short shower rather than a long, hot bath.

• Turn off the house lights, radio and television when they're not in use and keep the fridge and freezer doors closed.

• Write to government decision-makers and ask for laws that require people to use less energy. See page 89–91 for how.

CHAPTER 3:
WATER

Every time it rains, you are watching an important earthcycle at work—the water cycle. What happens if we waste or pollute water?

If we waste water by using too much, there may not be enough water to go around. Plants and animals that live in water habitats such as ponds, lakes, rivers and oceans, will be without homes. Wetland animals are already in trouble because their marsh homes are being drained.

Clean water is already scarce. In the 80 countries that contain nearly half of the world's people, water that is safe to drink is in short supply. In many other countries, people have to spend a fortune cleaning water before they can drink it.

Why is water unclean? Because of pollution. Chemicals turn rain water into acid rain, oil spills foul the oceans, and sewage and chemicals pour into our lakes and rivers.

Read on to find out where we get our water, what's happening to it, and how we can help to use it more wisely.

2. Clouds form. They are filled with water vapour. Eventually they get too full and the rain falls out.

1. Water from rivers, lakes and oceans evaporates into the air. Soil and plants give off water, too.

3. The falling rain seeps through the soil and eventually ends up back in a lake or river or the ocean. The cycle starts again.

49

Just add water

You are about 60% water, and no matter what you do about it, your body keeps leaking. You dribble out water in pee, you ooze water when you sweat, and your tongue and lungs dry out as you breathe. You have to keep adding water so that you don't shrivel up like a stale raisin.

You need about 1.25 L (1 quart) of liquids every day to replace the water you lose through peeing, sweating and breathing. Like a plant, you can't last long without regular watering. Half the water that you need comes from the food you eat. Surprisingly, much of what you eat is actually a drink. Some common foods are mostly water: eggs are 74% water, cucumbers are 95% water. Even bread is 35% water, but dry cereal is only 3% water.

The other half of your daily water needs comes from the stuff you drink. But finding a good clean drink is getting tougher. Where does your drinking water come from? Out of the tap? Ever wonder where your tapwater came from before it got to your house?

The water you drink was once a drop of rain, a speck in a stream or an eddy in a lake. It was blown by the wind, trickled over land, flowed into lakes or underground caverns and was eventually channelled into your water-supply system. But all the time it was out in the wild, your water was picking up extras. Bits of soil, other liquids, bubbles of air and animal and plant debris all get swept along into the water. Other less appetizing stuff also gets added to wild water. Chemicals from farm fertilizers and factories seep into the water. So do toxics from waste dumps and sewers.

Before you drink water from the tap, most of the worst pollution is removed and some good things are added. But even so, tapwater is never as clean as water that was never polluted. The only way to make it that clean is to get rid of the pollution *before* it enters the water, and that's a tall order.

51

Wasting wild water

How much water do you use every day? A pop-bottleful? A bucketful? A sinkful? A bathtubful? Believe it or not, the average American uses almost *three* bathtubs of water every day to wash, drink and flush. We use so much water that we're running out of it.

We don't need to use so much. Canadians use only two bathtubfuls, and Germans use less than one. And in some parts of Africa, people use less than one-tenth of the water we use. How can you use less? Let's look at your morning routine and see.

8:00 a.m.

Probably the first thing you do each morning is use the toilet. One flush uses 20 L (5 U.S. gallons) of water, enough to fill a kitchen sink twice. Toilet flushing accounts for nearly half of home water use in North America. To reduce your water use, see "A Full Flush."

A FULL FLUSH

To reduce your water use when you flush, all you need are some plastic bottles or jars filled with small stones. Use various sizes of bottles and jars. You're going to be putting them into the tank of your toilet, so they should be small enough to fit in.

1. Remove the top of your toilet tank and place it carefully on the floor.
2. Flush the toilet.
3. While it's re-filling, put a stone-filled jar or bottle into the tank. (Make sure the lids are securely closed on the bottles or jars before you put them in the tank.)
4. Flush again. Is there still enough water to carry waste away? If so, add another bottleful of stones to the tank before it fills up again. Repeat this step until the flush is too weak, then remove the last bottle. *Be very careful not to damage the flushing mechanism in the tank.*
5. If the plastic bottles in the tank have labels that tell you their volume, you can add up how much water you save each time you flush. Count the number of flushes per day for your toilet, and figure out your daily savings of water.

Leave the bottles of stones in your toilet tank and tell all your friends and neighbours to do the same. If every American saved flushing water this way, more than a billion litres (260 million U.S. gallons) of water would be saved every day — that's enough to fill 700 Olympic-size swimming pools each day!

8:15 a.m.

Time for a bath. A bath might feel great, but it's a big water waster. The average bathtub holds 150 L (40 U.S. gallons) of water. But if you take a shower, you use only a fraction of that amount. Want to find out how much water you can save by showering? Try "Shower vs. Bath" on this page.

SHOWER VS. BATH

How much water do you need to keep clean *and* environmentally virtuous? Here's one way to find out.

YOU'LL NEED:
a bathtub with a shower
a bucket
a 1 L (or 1 gallon) bottle
a watch
a scoop or a plastic bottle cut to make one

1. Use the litre (gallon) bottle to fill your bucket with water. Count to find out how many litres (gallons) of water the bucket holds.

2. At bathtime, fill your tub up to your usual level with bucketfuls of warm water. Keep track of how many bucketfuls you used. Multiply this by the number of litres (gallons) needed to fill the bucket, and you'll know how much water you used. How could you use less? Try bathing in a shallower tubful.

3. Next time you want to get clean, put the plug in the bathtub drain and take a shower. Time yourself. When you're done, scoop the water out of the tub into the bucket and count the bucketfuls as you pour them in the sink. Multiply the number of bucketfuls by the number of litres (gallons) in a bucket and you'll know how much water you used. Depending on your shower head and water pressure, a short shower might use between 20 L and 95 L (5 to 25 U.S. gallons). How do you compare? How many minutes did you shower? How could you use less water?

8:30 a.m.

Brush your teeth and try this test. Put the plug in the sink before you start brushing. Now turn on the water and keep it running as you brush your teeth. Note how high the water is in the sink. Let the water out. Now try the same experiment again, but this time only use the water to wet your toothbrush before brushing and to rinse it afterwards. Turn the water off in between. Not only are your teeth clean; you've saved a lot of water. (By the way, make sure to turn off the tap when you're done so that it doesn't drip. A dripping tap wastes 30 to 100 L [8 to 26 U.S. gallons] a day!)

You can save almost a whole bathtubful of water a day just by cutting your flushing, by showering instead of bathing and by turning off the tap between uses. And that's all before 9:00 a.m. Just think of how much more you could save in the rest of the day. Here are some other ways to save:

• Run a dishwasher only when you've got a full load, and save up to 35 L (9 U.S. gallons) a day.

• Run a washing machine only when you've got a full load, and save up to 60 L (15 U.S. gallons) a day.

• Water your lawn less. Watering the lawn for half an hour uses about 500 L (130 U.S. gallons).

WATER IT!

Water is also used to make many of the things we eat and wear. For example, watering crops uses about 73% of the world's water supply. How else is water used?

• *Producing a loaf of bread uses 544 L (145 U.S. gallons) of water.*

• *Producing cotton pyjamas uses 3600 L (950 U.S. gallons) of water.*

• *Producing a litre (quart) of milk uses 892 L (235 U.S. gallons) of water.*

• *Producing 0.5 kg (1 pound) of tomatoes uses 500 L (130 U.S. gallons) of water.*

• *Producing 0.5 kg (1 pound) of oranges takes 188 L (50 U.S. gallons) of water.*

• *Producing 0.5 kg (1 pound) potatoes takes 92 L (25 U.S. gallons) of water.*

Walking on water

Stand up and go outside. Congratulations! You're walking on water. Where's the water, you ask? Deep in the ground, under your feet. It's called groundwater, and it forms lakes and streams deep underground. These underground reservoirs are like huge tanks that hold the water until it is needed to supply stream, river and well water. Almost all of the fresh water in the world is found underground.

Groundwater used to be pure, but a lot of it is polluted nowadays. And there's another problem. In some parts of the world, fresh groundwater is being used up faster than it can be replaced. The result: water shortages. Eighty countries containing 40% of the world's population are short of water.

In Florida, fields are developing cracks and sinking in large chunks. Underground lakes and streams that used to be filled with groundwater are drying up—leaving air-filled caverns that cave in under the weight of the earth above them. What happened to all the water? It was used mostly to supply tapwater and to irrigate crops. In many parts of the world, overuse of water means water restrictions during dry spells. Some areas of California even have water police to make sure people aren't wasting valuable water.

Some regions have increased their supply of water by taking it from rivers in less populated parts of the country. For many years western states have had an agreement to channel water from the Colorado River. That leaves less for Mexico, where the river flows next. And what might kids in the future think about how we use rivers today?

Invisible pollution

Imagine taking a dip in a cool, clear brook. If it's a hot day, you might be tempted to take a drink of bubbling brook water. But don't. No matter how clean water looks, it's probably polluted.

Scientists test wild water carefully to see how polluted it is. You can, too. The tiny pieces of solid material in the water called "suspended particulate" are an important part of water pollution. Some of it may be harmless, and some may be poisonous, but in either case, too many particles can kill fish and insects and make swimming or boating unpleasant. Here's how you can test local wild water for suspended particulate.

YOU'LL NEED:
a piece of white paper or cardboard with clear black type on it
identical clean clear glass jars with clean lids
a felt pen

1. Hold the paper behind the jar and make sure that you can see the type clearly. Use only jars that allow you to do this.

2. Fill the jars with water samples from a nearby pond, ditch, lake or stream. Use a felt pen to mark on the top of your jar where the sample came from.

3. Shake each jar, then hold the paper behind it. Can you see the writing through the water? What about the dots on the i's, the commas and periods?

The more easily you can read the type, the fewer suspended particles are in the water and the cleaner it is.

To test for suspended particulate, the government scientists go out in boats or in hipwaders and lower a special card, called a "Secchi disc," beneath the water's surface. You can make your own Secchi disc by taking a circle of plastic or metal about the size of a saucer and painting it black and white like this:

Use a hammer and nail to make a hole in the middle. Thread a long string through the hole and tie a stone or other weight on the bottom side. How far can you lower your Secchi disc and still see the design?

The secret life of a glass of water

If you went to the lake or river where your tapwater comes from and scooped up a glass, you wouldn't want to drink it. Most likely, it contains a crazy cocktail (mixture) of chemicals, ranging from animal and human waste and dead organisms to a mess of nasty chemicals that have seeped, rained or flowed into the water. There's not much clean water left on our planet. Seventy percent of the earth is covered in water, but only 1% of it is salt-free and available for us to drink. Much of that 1% is now polluted.

Luckily for us, people have figured out how to clean up polluted or dirty water before they send it out into your tap. Here's how the water you drink may have been treated.

CHLORINE

WATER FROM RIVERS, LAKES OR WELLS IS SPRAYED INTO THE AIR OR PUMPED OVER RACKS TO EXPOSE IT TO THE AIR. THIS GETS RID OF SOME OF THE BAD SMELLS IN THE WATER.

IN SPECIAL RESERVOIRS, THE BIGGEST BITS OF SOLID MATERIAL DROP OUT ONTO THE BOTTOM. CHLORINE MAY BE ADDED TO KILL OFF THE GERMS.

If you live in the country and get your water from a well, or if you go to a cottage where the water comes from the local lake or river, your water might not be treated before it comes out of your taps. Want to know what you're drinking? Get in touch with your local environmental agency.

Raining vinegar

How is the rain that falls on some of Europe and North America like the vinegar you put on your salad? Both contain acid. Vinegar is acetic acid. Eaten in small amounts, it causes no harm. Acid rain, on the other hand, contains acid from air pollution. Although acid rain looks like normal rain, it can be incredibly harmful.

We humans make acid rain by using electricity, driving cars and operating heavy industries. The chemicals that spew out from car tailpipes, some factory smokestacks and power stations collect high up in the sky and are blown by the winds to faraway places. Eventually, the chemicals are caught in clouds and make the rain or snow that falls acidic, like weak vinegar. When acid rain and snow fall, they add acid to the soil and to the water.

The trouble is that acid, if it's strong enough, can poison animals and plants. (To see what an acid can do, try The Acid Test on the next page.) Acid can kill fish, prevent them from breeding or deform the young ones. It can kill or damage such plants as maple trees, tomatoes and pine trees that don't like acid soil. It can eat away at concrete and stone, destroying buildings, statues and monuments. Acid in the water can also release poisonous metals from the rock under lakes, streams and wells, further polluting the water and soil. If acid water flows through the water pipes in a house, it can absorb some of the metal in those pipes, too, making the tapwater unhealthy to drink. Acid in the air can hurt people, especially those with lung problems, making it harder for them to get the oxygen they need.

Acid rain is weaker or stronger, depending on the time of year and the wind direction. In northern countries, when the snow melts in the spring, all the acid that has been held in the snow all winter suddenly rushes into the streams and lakes. This "acid shock" can be disastrous for young fish, which are very vulnerable. A whole generation can be wiped out in a few days. This may mean fewer fish in the area for years to come. Most pollutants gradually build up over time, damaging the environment very slowly, so that it is often difficult to tell whether anything is being hurt at all. Acid shock happens fast—and it's deadly.

THE ACID TEST

You can find out for yourself what acid can do.

YOU'LL NEED:
*things to test, such as leaves, chicken bones, a hard-boiled
    egg with the shell on, or paper with printing on it
2 glass jars
water
vinegar
labels and pen or pencil*

1. Divide your test item into two parts and put one in each jar. For example, if you're testing eggs, use two, and put one in each jar.

2. Cover one item with water and the other with vinegar and label them so you know which is which.

3. Leave both jars for several hours or days, until you can see a difference.

4. Try testing other items.

What's the difference between the two jars? Pay special attention to the eggshell, chicken skin, print on the paper and the colour of the leaves. Acid rain is like weak vinegar. Imagine what it does to the objects it falls on.

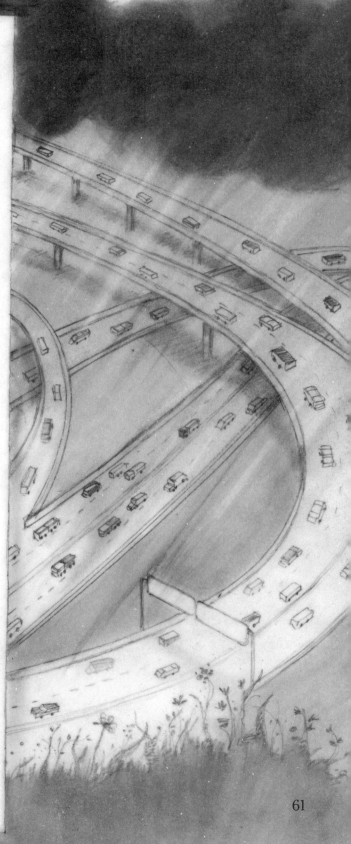

ACID TESTING

The amount of acid in any liquid can be measured by something called pH. The most acid materials have a pH of 0; the least acid would rate 14. Neutral is 7, and clean rainwater should measure 6 or so. Lakes filled with acid rain can dip as low as pH 4 – a disaster for the animals and plants living there.

Some common household items have very high or low pH values. You can measure their acidity with the help of some food.

YOU'LL NEED:
a knife
a red cabbage
a blender (or mortar and pestle)
a small glass for each material you want to test
test materials (try Coke, lemon juice, vinegar, baking soda, milk of magnesia and milk)

1. Use the knife to cut up the red cabbage, then put the cut-up cabbage in a blender with a little water. Blend until you have a mush. (If you don't have a blender, use a mortar and pestle and muscle power.) The mush you end up with is your acid detector.

2. Divide your acid detector among the glasses.
3. Add a bit of test material to the acid detector in one glass. Keep adding test material bit by bit until this acid detector changes colour. The more material needed to cause a colour change, the *less* acid the material is.
4. Add different test materials to different glasses. Which material causes a colour change most quickly? It is the most acid. See page 96 for the acid ratings of your test materials.

A POND IN A POT

When lakes get hurt by acid rain, they can sometimes be healed by adding stuff to the lake that takes away the acid. This stuff is a "base," the opposite of an acid. When you mix red paint with white paint, you get pink, a colour in between red and white. The same sort of thing happens when a base is added to an acid. The acid and base combine to form something in between that is neither an acid nor a base. This process is called "buffering," and it has been used to help lots of acid lakes so that they can once again support life.

You can make your own mini acid lake and then buffer it.

YOU'LL NEED:
water
vinegar or lemon juice
a cooking pot
an acid detector, like red cabbage (see Step 1 of
*   Acid Testing)*
dishwasher detergent
a cup

1. Put an equal mixture of water and vinegar (or lemon juice) into your pot.
2. Check the mixture with your acid detector. To do this, pour a bit of your acid water into the acid detector. Add more vinegar or lemon juice to the water if you need a lot of water to make your mush change colour. Now you have your own acid pond.

3. Mix a couple of spoonfuls of dishwasher detergent with water in a cup. This is your base. Pour this base into your acid pond and stir.
4. Use some acid detector to test how the pond is now. Is it still acid?

One of the most common bases is limestone. Some parts of the world have limestone already in the rocks under lakes and soil. This gives those lakes and lands a natural ability to buffer acid rain. For this reason, the same amount of acid rain can fall on two lakes, and one can end up dead and acidic, while the other, on limestone bedrock, can flourish.

Lakes found in areas with no limestone in the rocks under them are especially endangered by acid rain. The map shows the parts of the world that are most at risk from acid rain. Are your local lakes in trouble?

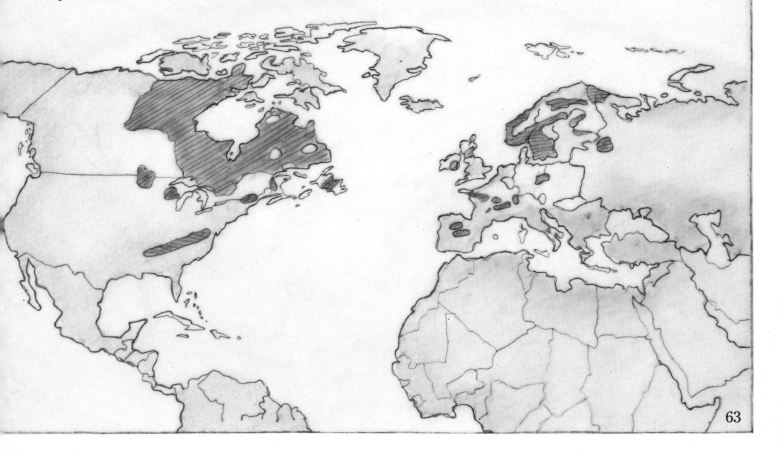

Oil on the ocean

Have you ever noticed a puddle in the street shimmering with rainbow colours? The same thing can happen on pavement where cars have been parked. This is a sign of oil or gasoline leaks. It's bad news in the city, but in the ocean it can be a killer. In fact, oil leaks are one of the most serious forms of ocean pollution.

Up to 6 million tonnes (tons) of oil are dumped into the world's oceans every year. Oil seeps out of boats that use it as fuel or carry it from port to port, and it flows into the sea from factories on the shore. Oil does much the same thing in the sea that it does in a puddle: it spreads out in a thin layer on the surface of the water.

A large oil slick can cover many square kilometres (miles) of ocean, and it can last for up to ten years. It prevents the water from absorbing bubbles of air and, as a result, cuts down on the oxygen below the surface that fish need to survive. Diving birds can be blinded by the spilled oil, and other animals that live on or near the surface of the water can be covered in it, too. When birds try to clean it off their feathers, they swallow some of the oil and may be poisoned. People who try to help out by washing the oil off the birds can destroy the birds' natural waterproof covering so that the birds get waterlogged and drown. If detergent is added to the oil to break it up and make clean-up easier, it often kills the fish and other marine life.

CREATE AND CLEAN YOUR OWN OIL SLICK

You can find out what an oil slick is like by making your own.

YOU'LL NEED:
cooking oil
a cup
paprika or turmeric
a spoon
a bowl of water
marbles, feathers, small pieces of paper or paper clips
a large popsicle stick or spatula
cotton wool
dishwashing detergent

1. Put a couple of big spoonfuls of cooking oil in a cup. Add the paprika or turmeric to colour it, and use a spoon to mix it well.
2. Pour this coloured oil into the bowl of water. Now you have an oil slick.
3. Drop objects such as a marble, feather, piece of paper or paper clip into the water. Remove them and what do you see? They are coated with a thin film of oil, just as waterbirds are in a real oil spill.

4. Now try cleaning up your oil spill. In real life, special emergency clean-up crews rush to the site of an oil spill and try to clean it up, or at least collect as much of the oily film as possible. Use a large popsicle stick or spatula to skim off the oil from the surface of the water. Clean-up crews at oil spills use this technique. They float booms, like large logs, around the spill to contain it and then skim off the oil.

5. Try absorbing the oil with cotton wool placed around the edge of your bowl at the waterline. At a real oil spill, clean-up crews sometimes place straw on the edge of the shore to absorb the oil.

6. Detergent-like chemicals may be added to break up the slick into small drops. Try adding a spoonful of detergent to *your* oil spill. Beads of oil are easier to clean up on beaches, but detergent adds other pollutants to the sea.

Recently huge oil spills have gushed into the sea in the Middle East and off the Alaska coast. Many animals are killed during an oil spill.

If you think this is bad, think twice before you ride in a car. A 1989 oil spill in Prince William Sound, Alaska, sent 267 000 barrels of oil gushing into the ocean. American cars leak more than 50 times that much oil every year! To reduce oil leaks from cars, walk or bicycle instead of asking to be driven in a car.

How else can you use less oil? A great deal of oil in North America is used by industry, to power factories, to heat buildings and to make products such as plastics. You can't do much on your own to reduce oil used in these ways. But you can make your concerns known to government. See pages 88–93 for how best to do this.

Overfed lakes

When soil is washed or blown from farm fields into rivers and lakes, the fertilizers in the soil go, too. These fertilizers make the water more nutritious for the plants. Sound good? Unfortunately, water plants fed by fertilizers grow too profusely and choke out other water creatures. See for yourself.

YOU'LL NEED:
plant fertilizer or fish food
a glass jar or fishbowl full of water, with water plants if possible but without *fish*

1. Add lots of fish food or plant fertilizer to your bowl or jar.
2. Place it in a sunny spot and wait for a week or two, adding more fertilizer or food after a week.
3. What's in your bowl now? Is it greenish? Is there a scummy skin on the water? If so, you're growing algae, and your bowl is no good as a home for most fish.

From food to fertilizer

You have a recycling plant in your body. It takes food and turns it into solid waste. This solid waste is full of nutrients you didn't digest. It could be purified and used to fertilize crops. But it rarely is.

Instead, human waste, called "sewage," often finds its way into the lakes and rivers and oceans and fertilizes them. The result: water plants grow and choke out the animals. In the last few years, governments have passed laws to improve sewage treatment so that human waste doesn't end up overfeeding lakes and rivers. But some cities and countries still dump their sewage right into the water. Some U.S. cities have built long pipes to carry sewage miles away from shore and then spew it into the ocean. Does this bug you? See Raise a Ruckus *on page 88 for what you can do.*

The same thing happened to many lakes in the 1970s. Farm fertilizers and detergents rich in nutrients were washed into lakes. The lakes became murkier as lots of scummy blue-green algae grew in them. The decaying algae used up most of the oxygen in the lakes that fish need to live. Eventually, many types of fish started to die off in these lakes. Fish (such as lake trout) that need lots of oxygen were replaced by other kinds of fish (such as carp) that could survive in these oxygen-starved, or "eutrophic," lakes. Fishermen were mad; they preferred lake trout and other fish that need lots of oxygen.

So governments brought in laws to limit the amount of nutrients in products such as detergents to prevent the over-feeding of lakes. Once-murky lakes such as Lake Erie were cleaned up, and many of the fish that used to live in them flourished there once more.

Down at the mouth

Fertilizer isn't the only thing that gets washed or blown into the water. Pesticides are washed off farmers' fields and off people's lawns and gardens, poisons in car exhaust are swept from roadways, and industrial chemicals are flushed away from factory yards. When these chemicals end up in a river, they are carried down to the river's mouth. There the water travels too slowly to move them any farther.

These poisons stay buried, until they are disturbed. When the mouth of the river gets so clogged up that ships passing through are in danger of running aground, special dredges scoop up the muck and, in the process, uncover the nasty chemicals buried in it. These chemicals can spread through the water, poisoning wildlife and sometimes even getting into our water supply.

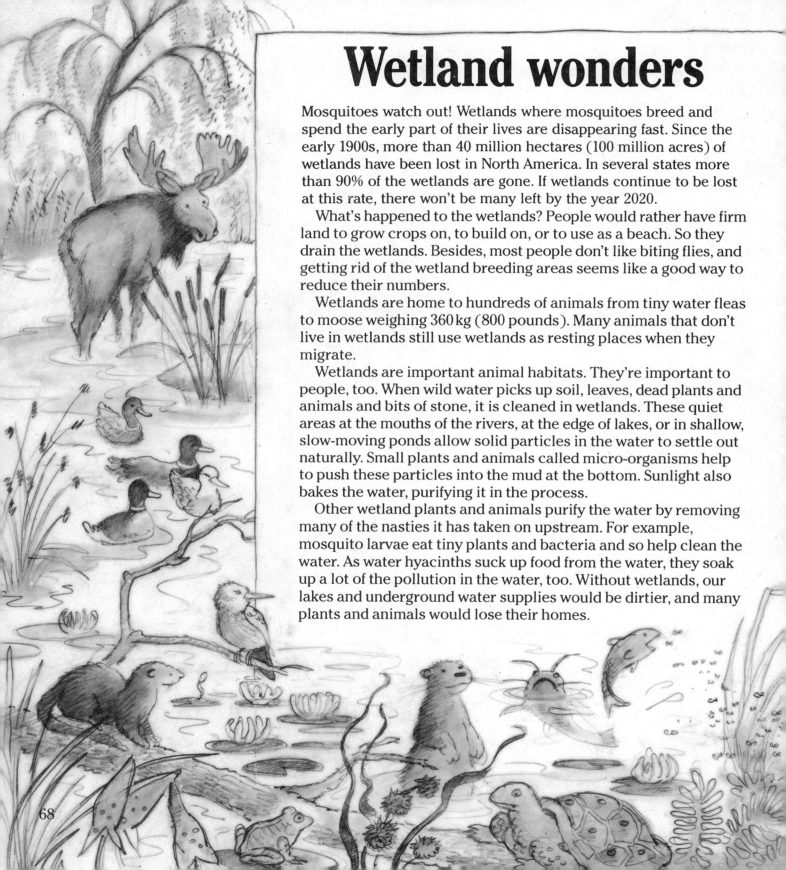

Wetland wonders

Mosquitoes watch out! Wetlands where mosquitoes breed and spend the early part of their lives are disappearing fast. Since the early 1900s, more than 40 million hectares (100 million acres) of wetlands have been lost in North America. In several states more than 90% of the wetlands are gone. If wetlands continue to be lost at this rate, there won't be many left by the year 2020.

What's happened to the wetlands? People would rather have firm land to grow crops on, to build on, or to use as a beach. So they drain the wetlands. Besides, most people don't like biting flies, and getting rid of the wetland breeding areas seems like a good way to reduce their numbers.

Wetlands are home to hundreds of animals from tiny water fleas to moose weighing 360 kg (800 pounds). Many animals that don't live in wetlands still use wetlands as resting places when they migrate.

Wetlands are important animal habitats. They're important to people, too. When wild water picks up soil, leaves, dead plants and animals and bits of stone, it is cleaned in wetlands. These quiet areas at the mouths of the rivers, at the edge of lakes, or in shallow, slow-moving ponds allow solid particles in the water to settle out naturally. Small plants and animals called micro-organisms help to push these particles into the mud at the bottom. Sunlight also bakes the water, purifying it in the process.

Other wetland plants and animals purify the water by removing many of the nasties it has taken on upstream. For example, mosquito larvae eat tiny plants and bacteria and so help clean the water. As water hyacinths suck up food from the water, they soak up a lot of the pollution in the water, too. Without wetlands, our lakes and underground water supplies would be dirtier, and many plants and animals would lose their homes.

Natural early warning systems

One wetland dweller, the dragonfly, acts as an early warning system for pollution. Dragonflies are one of the first animals to be hurt by pond pollution. When the dragonflies at a pond disappear, other pond plants and animals will usually follow, if the pond isn't cleaned up.

Dragonflies are a real asset to people, too. They are big eaters, and one of their favourite foods is mosquitoes. The largest dragonflies can eat many dozens of mosquitoes a day. What's more, they eat the mosquito at every stage of its development. Wells, Maine, has even used dragonflies to combat their mosquito problem. They found that dragonflies were cheaper and safer than spraying pesticides. Using natural methods for pest control is called "biological control." When it works, it's a lot better for us and for the environment than using chemicals.

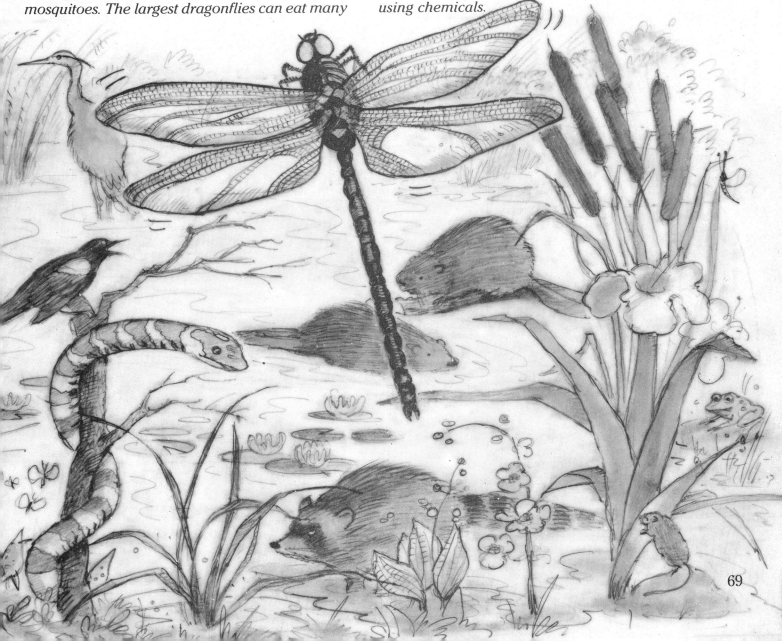

69

Poison in the parlour

There are killers lurking in your family's cleaning cupboard. Take a look and see. Do any bottles or cans have pictures like these on them?

These are all signs warning you to handle these cleaners with care because they can be harmful to your health. Cleaners that are bad for you will also be bad for lakes, rivers or oceans, which is where they'll end up if you pour them down the drain.

How much danger do you pour down the drain in your home? All you need is a felt-tip marker to find out.

1. Look in your cleaning cupboard and find two common household cleaners that have warning labels. Use your marker to mark the level of the liquid or powder on each.

2. A month later, mark the new level for each cleaning material. How much was used? This is how much cleaner you've used in a month.

Think about the huge amount of cleaner that gets dumped down the drains of North American homes every month. Over the years, this adds up to lakefuls of the stuff, which pollutes our lakes, rivers and seas. You can help your family reduce the amount of toxic wastes you pour down the drain. Make-your-own, environment-friendly cleaners do a great job, and they do much less harm while they're at it. Try these recipes:

• Mix together equal parts of ammonia, vinegar and baking soda and add 15 parts water to clean floors.

• To polish furniture, try rubbing on mayonnaise with a soft cloth, or use a mixture of two parts vegetable oil and one part lemon juice.

• To clean copper pots and jewellery, scrub with vinegar and salt.

Warning! Never mix chlorine bleach with ammonia or vinegar or toilet bowl cleaners—these can produce poisons!

Be a toxic cop

Can you spot 14 things that shouldn't be poured down the sink or put out in the garbage? (Answers on page 96.)

CHAPTER 4:
SOIL

What happens when a plant or animal dies? It decays and adds nutrients to the soil so that new plants can grow. It becomes part of the soil cycle.

The land not only gives us food—it also gives us paper, lumber, glass, metals and many other things. Look around you. Everything you see started in the soil or rocks.

Land is precious. That's why the native people believe that the land belongs to everyone. But we don't always treat the land well. We waste the paper, food and metals that are provided by the land, so that it must produce more and more for us. Meanwhile, the stuff we waste ends up in garbage dumps that take up more and more valuable land.

4. Animals die and decay, adding rich nutrients to the soil.

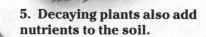

5. Decaying plants also add nutrients to the soil.

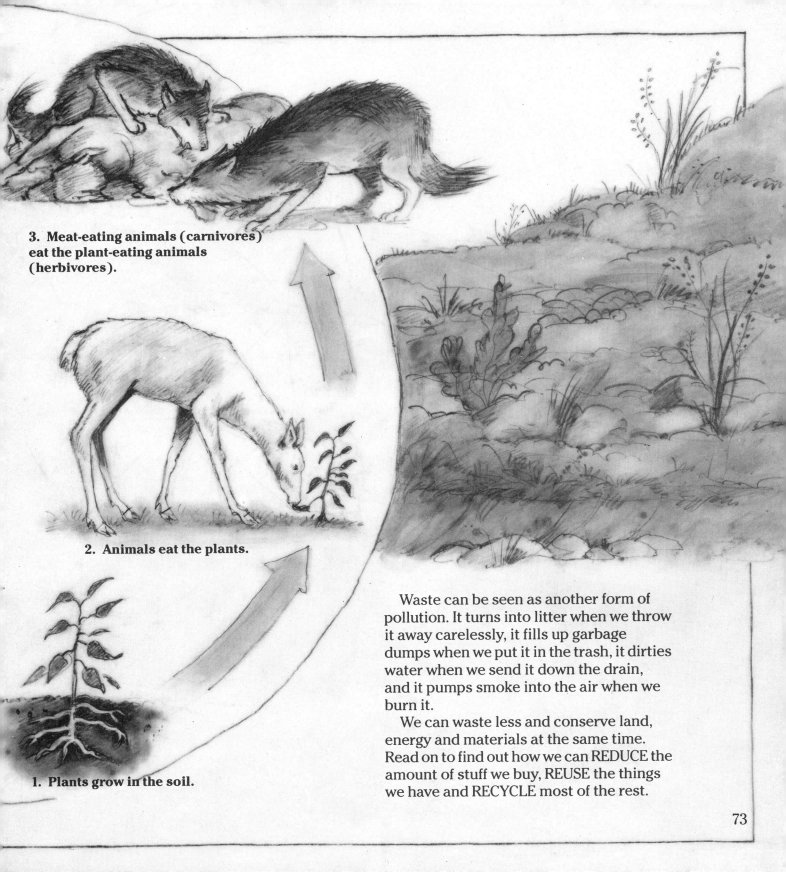

3. Meat-eating animals (carnivores) eat the plant-eating animals (herbivores).

2. Animals eat the plants.

1. Plants grow in the soil.

Waste can be seen as another form of pollution. It turns into litter when we throw it away carelessly, it fills up garbage dumps when we put it in the trash, it dirties water when we send it down the drain, and it pumps smoke into the air when we burn it.

We can waste less and conserve land, energy and materials at the same time. Read on to find out how we can REDUCE the amount of stuff we buy, REUSE the things we have and RECYCLE most of the rest.

73

Eat dirt!

Have you heard the expression "You are what you eat"? If so, you must be mostly dirt. You don't eat dirt, you say? Maybe not directly, but everything you eat comes from the soil. Plants grow in soil and end up on your table. Animals eat plants that grow in soil, and the *animals* end up as meat you eat. So the vitamins and minerals you eat for lunch used to be in the earth before they were absorbed by plants, roots, turned into vegetables, and eaten by you or the animal that produced the meat in your sandwich.

It takes a lot more land to grow meat than to grow vegetables, fruit or grains. In fact, it takes about 100 times more land to produce beef or lamb than to grow tomatoes or potatoes. So when you put meat on your table, you're using lots of land to do it.

How much land does it take to grow the foods you eat in a year? Try this and see:

Burgers and Buns

How much land does it take to produce the hamburgers you eat in a year? Get out a calculator and find out.

Calculate how many hamburgers you eat in a year. If you eat 8 a month, multiply 8 by 12 months.

$$8 \times 12 = 96 \text{ hamburgers a year}$$

For the hamburger meat:

It takes 20 square metres (20 m^2) of the very best land to grow the meat in one hamburger. That's as big as the floor of a living room.

How much land does it take to grow all the meat in the hamburgers you eat in a year?

$$96 \times 20 \text{ m}^2 = 1920 \text{ m}^2$$

For the buns:

It takes 0.2 m^2 to grow the wheat in one hamburger bun. That's about the half the size of a school desk.

To grow all the buns you eat in a year it would take:

$$96 \times 0.2 \text{ m}^2 = 19.2 \text{ m}^2$$

So it takes 100 times more land to grow the meat than the bun. Do some simple addition to see the total amount of land you need to grow all the hamburgers you eat in a year:

$$1920 \text{ m}^2 \text{ of land for the meat}$$
$$+ \quad 19.2 \text{ m}^2 \text{ of land for the buns}$$
$$1939.2 \text{ m}^2 \text{ of land in total}$$

This is a chunk of land bigger than a hockey rink. On poorer cattle pasture land, you'd need a chunk of land bigger than a football field—just to grow your hamburgers. Imagine how much land it would take to grow all the food you eat!

Prehistoric meal plan

People are really just sophisticated animals. In prehistoric times, people didn't grow things on farms; like other animals, they just found or hunted food and ate it raw. Here's what someone might have eaten in one day, 5000 years ago:

- *sparrows and small mammals such as rodents*

- *seeds*

- *fish*

- *acorns, butternuts, hickory nuts and walnuts*

How does this compare to your daily meals?

Going, going, gone

Have you ever tried to wash away the sand or dirt in the bottom of the sink with a strong spray of water from the taps? Nature does the same thing each spring. Water from melting snow sweeps earth along with it into streams and rivers. On land covered by forests or fields, the roots of the trees and grasses keep a firm grip on most of the soil. But when trees are cut and fields are cleared of plants, more of the soil is washed away.

Wind steals valuable soil, too. It sweeps over bare areas drained of water and cleared of plants, carrying dirt with it as it goes. Huge dust storms can strip most of the good growing soil bare.

This is called "erosion," and it sweeps more than 25 billion tonnes (tons) of soil into oceans and rivers by wind or water each year.

SAVE YOUR SOIL!

You can find out how to create or prevent erosion by trying to flush away a mini mountain.

YOU'LL NEED:
a bucketful of sand or earth
a sidewalk, near a lawn or garden, to absorb the water and sand
a hose or a bucketful of water
gravel, old rags, grass sod or potted plants

1. Build your sand or earth into a mini mountain on the sidewalk.
2. Using your hose or water bucket, try to wash away (erode) your mountain.
3. Rebuild your mountain and try to stop the erosion by protecting it with a rag, gravel or plants. (Put the pots as close together as you can, and shove them down into the earth or sand.)
4. Spray or slosh the water onto the mound and see if the earth flushes away now. Did you slow down the erosion? (Don't forget to clear up your sand when you're finished, so you don't block the sidewalk.)

Disappearing farmland

Erosion blows away topsoil—the good rich soil that farmers depend on. But farms face another threat. Farmland is being paved over and dug up at an alarming rate to make room for new houses, offices and factories. Cities are gobbling up precious farmland that is needed to grow food.

How the earthworm ate the robin

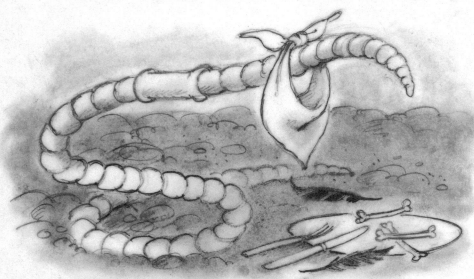

You've seen a robin eat an earthworm, but have you ever seen an earthworm eat a robin? Before you say no, think of this: when a robin dies, its body decomposes (rots) and becomes part of the soil. The earthworm eats the rotting material in the soil and so consumes the robin. Earthworms are great recyclers.

Even animals such as hawks and wolves, which aren't usually killed by other animals for food, die eventually, and their remains become food for worms and tiny organisms. These dead-meat-eaters, called "decomposers," turn carcasses into soil by grinding up the material into tiny particles.

Thanks to these garbage gobblers, nothing gets wasted.

Rotting plants and animals get turned into nutrient-rich soil for other plants to grow in. The only garbage that doesn't get recycled in nature is the trash we humans leave behind. Everything else that is left by one animal or plant gets used by another.

Without all these garbage gobblers, mountains of dead leaves, fallen trees and carcasses of dead animals would pile up everywhere. Plants would suffer. The soil would become poorer and poorer because it would lack the nutrients that come from decomposed plants and animals. Animals would suffer, too. Mice and other mammals that live on the forest floor would lose their homes under the piles of garbage and die.

Without these animals to eat, predators would starve. The entire forest community would suffocate or starve, buried under its own garbage.

Natural garbage gobblers help keep our human garbage under control, too. Check out how well they can deal with your family's garbage.

YOU'LL NEED:
a trowel or large spoon
several containers full of earth or a back yard where you can dig holes
different kinds of garbage, such as a disposable diaper, a plastic shopping bag, eggshells, orange peel, meat or fish bones, gristle, bean stalks or cauliflower leaves
labels (pieces of paper stuck onto toothpicks or Popsicle sticks work well)

1. Use your trowel or spoon to dig holes in the soil. Put one piece of garbage into each. Cover with earth. Label each buried thing, so you know what's where.

2. Water your garbage lightly twice a week.
3. After two weeks, dig up your buried garbage. How different does it look now?
4. Rebury your garbage, and dig it up again after another two weeks. What kind of garbage rots first? Why?
5. Try this experiment at different times of the year. Do garbage gobblers work better when it's warm or when it's cold? When it's wet or when it's dry?

Living chain

All animals can be divided into groups, based on what they eat:
- *Carnivores eat meat (other animals).*
- *Herbivores eat plants and eat no meat.*
- *Omnivores like us eat a bit of everything.*
- *Decomposers, such as cockroaches and worms, are nature's garbage cans. They eat any animal or vegetable material that's left over from other animals' needs.*

Put them all together and they form a kind of living chain. Here's one chain:

the hawk eats the bird

which eats the worm

which eats the plant

Add more animals that eat worms or birds and you have a web instead of a chain. Food chains and food webs link the community together.

How much earth do you throw out?

Are you throwing out valuable soil? You are if you don't compost. A composter turns kitchen scraps into rich new soil for your garden or houseplants. One-third of the stuff you leave out on garbage day could be put into a composter. You'd have healthier plants, and your local dump wouldn't get so full so fast.

You can see what a difference composting makes to your garbage output by trying this:

YOU'LL NEED:
a pen and note paper
small-sized garbage bags (not the huge ones)
composter
the co-operation of your family for a month!

1. Count the number of garbage bags your family fills in two weeks.
2. For the *next* two weeks, put everything you can into the composter (for how to make a composter and what to "feed" it, see the next page). Count the number of garbage bags you fill with the remaining garbage.
3. Do some simple math. Take the number of bags from the first two weeks and subtract the number of bags of garbage from the second two weeks. The number you end up with is how many bags of garbage you saved by composting. This not only benefits your plants; it also saves space in your local dump.

The average American produces more than 100 garbage cans full of garbage every year. Composting can cut that to as few as 70 cans.

Not everyone is as wasteful as we are. Austrians throw out only about ten cans full a year. In India, each person produces only about nine full cans.

MAKE A COMPOSTER

YOU'LL NEED:

a wooden packing or orange crate with holes drilled in the sides
a large plastic garbage bag
soil
kitchen garbage, including fruit and vegetables, coffee grounds, tea leaves, eggshells, and/or garden waste such as grass clippings. Don't include milk, cheese, yogurt, meat or fish. These will attract animals. And don't include foil, glass, metal or plastic, which don't decompose.
a thermometer
a pitchfork or spade

1. Your wooden crate will serve as your composter. Put it outside in the garden or on a balcony. (For balconies, set the composter in a low-sided box lined with plastic. Your composter may make some mush, and you don't want it to leak all over the balcony.)

2. Now you need to expose your compost to soil and the organisms it contains so that they can break down your left-over food and turn it into rich earth. Put your left-overs into the composter and cover them with a layer of soil.

3. Keep your composter wet but not soaked. Water it every week, if it doesn't rain. If it rains too much, cover the top with a plastic garbage bag.

4. After two weeks, test the temperature of your pile by shoving the thermometer into the middle of it. Wait for three minutes, then check on the reading. Your pile should be about 60-70°C (140-160°F). If the temperature is lower, try adding more soil or water, or stir the compost up to add air. When the garbage is converted to soil, its temperature should drop to 40-50°C (100-200°F).

5. Keep doing Steps 3 and 4 for a few months, checking the temperature and turning the whole pile with a spade or pitchfork every month. At the end of four months, you'll have some rich fertile soil, where once there was only garbage. Use your new earth to perk up potted plants, fertilize your vegetable garden, or sprinkle on your flowerbeds.

All wrapped up and nowhere to go

What do you bring home from the supermarket? Food? Soap and other cleaning materials? Bathroom supplies? You bring home all these things, plus some other stuff that you don't need, stuff that ends up in the garbage. What is this mysterious stuff? Containers, cardboard, wrapping and other packaging.

Packages do serve some important purposes. They give you information and instructions about the content and use of your purchases. They tell you the price, and they keep your food clean and protected from getting scrunched or contaminated with germs. But too much of your weekly garbage is composed of this material. You can see for yourself how much packaging you

bring home by collecting all the packaging your family throws out for a week. Use separate garbage bags and throw in anything you don't eat or use—boxes, cans, jars, cardboard, bubblepacks (you've probably seen this kind of packaging if you've bought batteries for a game or toy), bottles, plastic tubs and bags.

On average, about one-third of the garbage you produce is packaging. If every North American family could avoid packaging completely, there would be nearly 10 million fewer garbage cans full of garbage each year in dumps. That would not only save dump space, it would also save all the paper, plastics, energy and other resources used to produce the packaging in the first place.

How to reduce your waste

Here are some tips for cutting down on your packaging:

- Take your own shopping bags to the store.

- Buy in bulk, putting loose cookies, rice, etc., into bags rather than getting the pre-packaged variety in bulky containers.

- Buy in condensed form; for example, choose frozen rather than pre-mixed juice.

- Buy beverages in returnable bottles whenever you can.

- Choose large containers rather than several small ones.

- Buy vegetables individually, rather than in pre-packaged lots with cardboard or plastic packaging.

Getting rid of garbage is a big problem — no one wants it, and we're running out of space in our garbage dumps. Some cities have become so desperate to stash their trash that they ship their garbage out to sea. Italian garbage has been known to travel the oceans for up to a year before being dumped in another country. A barge of New York trash cruised the Caribbean for 155 days and finally docked back in the U.S.A.

Some people have suggested that we should burn garbage to create power rather than bury it. But where there's fire, there's smoke, and the smoke from burning garbage often contains harmful chemicals. Burning garbage also destroys useful raw materials that could be recycled.

83

Buried treasure

Does the garbage dump on the left look different from the scene on the right? Look closely. Many of the things in this kid's room are made from used stuff that might have ended up in the dump. The address book beside the telephone, for example, is made from the picture postcards and scrap paper thrown out in the garbage. How many more treasures can you match up with the trash in the dump?

Reusing stuff that you would otherwise throw out is a good way to reduce waste and save money. Bottles, jars and yogurt containers can be reused to package food and drink that you buy in bulk. Leaflets and brochures that have been used on only one side make great drawing paper. Let your imagination run wild and dream up some other good uses for your garbage. Here are some ideas to get you going:

- Turn old milk and juice cartons into blocks, toy trains and flower or plant pots.
- Enclose cards or special pictures in used plastic bags. Bind them together to make covers for photo albums or journals.
- String together a hammock from plastic six-pack carriers (lots of them) and heavy string. (Six-pack carriers are dangerous for land and water animals, who can get their necks stuck in the holes and suffocate or starve.)
- Transform old gloves and socks into puppets.
- Wash out plastic bags and use them as freezer or lunch bags. (Keep them right side out so no ink rubs onto your food.)
- Cut cereal cartons down to make boxes for art paper, letters and magazines.
- Wrap presents in scraps of left-over fabric or wallpaper.
- Store bulk foods in old jam jars.

Gold in the garbage

Garbage is stuff that's no good to you any more. That doesn't mean it's no good to anybody. In fact, some people will actually pay for things that you throw out.

Many cities are trying to cash in their trash by recycling valuable materials in garbage and selling them to businesses who turn junk into new products. Some materials, such as aluminum pop cans, can be melted down and reshaped into new pop cans. Other products can be mushed up, reprocessed and transformed into something completely different. For example, plastic containers can be turned into fibre stuffing for pillows and jackets and into solid plastic for paint-brush handles and fence posts. You will sometimes see special labels on things you buy to tell you that they have been recycled.

Many cities now pick up old tins, bottles, newspapers and some plastics for recycling. How much of your family garbage do you recycle? Find out by following Steps 1-3 on page 80 but substituting recyclables for the compostables. Depending on your neighbourhood's recycling program, you may be able to separate and recycle up to one-third of your garbage (and that doesn't count the stuff you can compost!). Your family's recycling effort could save your town a cubic metre (1.3 cubic yards) of dump space a year—and that adds up quickly if everyone participates.

One of the most valuable things that we throw out is paper. More than one-third of our garbage is paper products, including newspapers and cardboard. On average, each American throws out about 260 kg (580 pounds) of paper every year! Many families already recycle newspaper, but fine paper (like writing paper and computer paper) is a bit more difficult to separate from the rest of our rubbish. You can help your family to reduce rubbish, save trees and stop land from being used for garbage dumps by setting up your own fine paper recycling program.

RECYCLE FINE PAPER

1. Call your local environmental agency to find out the names of local businesses or offices that recycle their paper. Visit them and ask if they will allow you to add your family's fine paper to their supply for the recycler. If they agree, get a list of what kinds of paper they will accept (not all kinds of paper can be recycled).

2. Become the family paper snoop. Collect recyclable fine paper, take it to the local business or institution that will recycle it with their own.

If you're feeling really energetic and have lots of friends or neighbours who want to help out, you can expand this program. Make copies of the list of what kind of paper is and isn't acceptable for recycling and give one to everybody who wants to participate. Check that your local business can still handle all the paper you plan to collect—you may need a separate collection if you get too big. Set up a central storage area for all the recyclable fine paper and get everyone to bring paper to this spot. Good luck!

87

Raise a ruckus

How do you feel after hearing some environmental horror stories? Angry, frustrated, depressed or indignant? Want to feel better—and help the environment too? Get involved. Do something to teach people about the environment, or to reduce pollution, protect endangered species or habitats, or conserve resources.

How can you best make a difference? Many environmental problems are just too big for any one person to solve on his or her own. But if we work together, we can clean things up.

The first task is to persuade other people to care about the environment as much as you do. Some of the most important people to persuade are the politicians who can pass laws that will harm or help the environment. See the box on this page for a list of who to write to.

A good way to tell politicians what you think is to write it down and put it in the mail. Most politicians figure that one letter received means that up to 20 other people feel the same way. So your letter carries a lot more weight than you might think. Remember—politicians care a lot about what voters think; they want to get re-elected. The same thing is true for the people in charge of big companies: they need the support of the people out there buying their products. So let them know if you think the stuff they make is produced in a way that creates pollution, garbage or waste.

The sample letter on the next page has some attention-getting tips to get you started.

Whom to write to

These people make decisions about the environment. Send letters to them and make your concerns known.

President of the U.S.A.

Secretary of the Interior and Director of the Environmental Protection Agency

Secretaries of Commerce, Transportation, Energy, and State, depending on the issue

Representatives and Senators

Your Governor and legislators

Mayors and city government

89

Flattery will get you somewhere.

Praise positive actions taken in the past.

Be as specific as possible.

Ask for a response.

Get friends to include their signatures. The more signatures, the greater the impact.

This used to mean "carbon copy." It tells the politician that you're sending copies of your letter to other people, including the local paper.

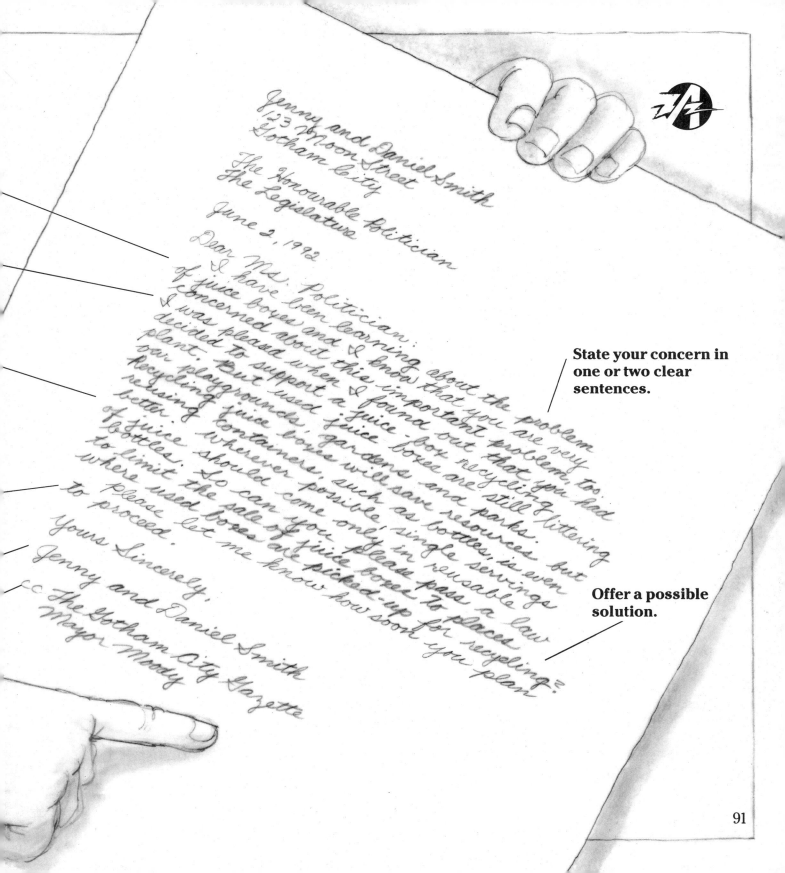

State your concern in one or two clear sentences.

Offer a possible solution.

Roll up  your sleeves

Here are some other ideas for things that you and your friends can do to help the environment:

Create environmental T-shirts by writing environmental messages onto plain T-shirts and wearing or selling them. Donate the money to your local environmental group.

Talk to your parents, friends and relatives who are old enough to vote. Persuade them to vote for candidates who care about the environment.

Circulate petitions about issues that make you mad. Send them to environmental decision makers.

Put together environmental posters or newsletters for your neighbours, telling them about environmental problems and how they can help.

Write letters to the editor of your local newspaper and call phone-in shows on the radio, explaining your environmental worries and what you want done about them. If you don't get a response within a few weeks, write or call to follow up.

Write a play or a puppet show on the environment and present it to other children or to adults.

Find out how you can help local environmental groups. They usually have more work than they can handle.

GLOSSARY

Acid rain   Air pollution (sulphur dioxide and nitrogen oxides) that combines with water vapour in the clouds, and makes the rain and snow acidic

Biological control   The use of natural, instead of chemical, methods to control pests. These methods include using natural predators and disease-causing viruses and bacteria.

Buffering   Adding a base to an acid to neutralize the acid. Buffering is used to help restore lakes that have been damaged by acid rain.

Carbon dioxide   A colourless, odourless gas created by breathing and by burning coal and other fossil fuels. It gets used up by green plants in photosynthesis.

Carbon monoxide   A harmful chemical produced mainly by cars. It is one of the major air pollutants.

Clear-cutting   Cutting down all the trees in an area, no matter what their size, shape, age or quality

Chlorofluorocarbons (CFCs)   Gases that destroy the ozone layer and contribute to global warming. They are manufactured by people and are found in refrigerators, air conditioners, and some spray cans and Styrofoam packaging.

Chlorophyll   The chemical that gives green plants their colour. It is essential to them for making food.

Decomposers   Animals such as worms and cockroaches that consume animal or vegetable remains and turn them into nutrient-rich soil

Ecosystem   The relationships between plant and animal species and their environment, in a natural community

Endangered   Will become extinct if threatening conditions are not changed quickly. An endangered species may have too few members left to carry on the species.

Erosion   The wearing away of the Earth's soil by wind and water

Extinct   None left in the world. A species is extinct when all its members are gone.

Extirpated   Gone from most former habitats and now found only in a few places

Greenhouse effect   The heating up of the Earth caused by rays of the sun trapped within the Earth's atmosphere by carbon dioxide

Habitat   The physical surroundings of an organism. The habitat provides the organism with everything it needs to live.

Nitrogen   A colourless, tasteless, odourless gas which makes up about 4/5 of the atmosphere

Nutrients   The food or foods a plant or animal needs to survive

Oxygen   A colourless, tasteless, odourless gas, essential to all life and found in most things

Ozone layer   A layer of ozone (a type of oxygen) that prevents most of the sun's dangerous rays from reaching the Earth

Pesticides   A chemical substance made by people that is used to kill insects. Pesticides can damage ecosystems, and often harm wildlife other than the insects they are intended to destroy.

pH   The unit of measure of acidity, which ranges from 0 (most acid) to 14 (basic or alkaline). Neutral is 7.

Photosynthesis   The process that occurs when plants use chlorophyll and sunlight to convert water, carbon dioxide and nutrients into food

Rainforest   Lush, wet forests that cover 6% of the Earth's land area and provide a large amount of the world's oxygen supply. They contain half of all plant and animal species in the world.

Regeneration   Redevelopment of an ecosystem (such as a forest) after it has been partially or completely destroyed

Toxic   Poisonous

Wetlands   Quiet pools of water found at the mouth of rivers or the edge of lakes. Wetlands are home to many plants and animals.

INDEX

ANSWERS

Just add water, pages 50–51:
Did you spot these sources of pollution: farm fields being sprayed with chemical fertilizer, the smokestack spewing dirty smoke, cars going over a bridge, left-over waste from someone washing his car flowing into the storm sewer grating on the street, a deforested area with soil being blown or rained into the water, leaking septic tank (In the country, most people don't use sewers. Instead, the waste from their toilets is treated on the spot. However, sometimes the septic tanks don't work as well as they should), leaking toxic chemical drums, the factory dumping chemicals, power boat leaking gas.

Acid testing, page 62:
Here are the pH numbers for each of your test materials (for a description of pH see page 62). The lower the number, the *more* acid in the test material.

Lemon juice and vinegar	2
Coke	2.8
Milk	6.6
Baking soda	8
Milk of Magnesia	11

Be a toxic cop, page 71:
Did you spot these toxic chemicals: antifreeze, batteries, bleach, bug spray, glue, old medicine and prescriptions, oven cleaner, paint, paint remover, polish remover, pool chemicals, rubber cement, rust remover, varnishes. These are all too dangerous to be put out in your regular garbage. Call your local garbage collection department to find out how to dispose of them safely.